Hochschulschriften

Institut für Systembiotechnologie
Universität des Saarlandes

Herausgegeben von Prof. Dr. Christoph Wittmann

Band 6

Cuvillier-Verlag
Göttingen, Deutschland

Herausgeber
Univ.-Prof. Dr. Christoph Wittmann
Institut für Systembiotechnologie
Universität des Saarlandes
Campus A1.5, 66123 Saarbrücken
www.iSBio.de

Hinweis: Obgleich alle Anstrengungen unternommen wurden, um richtige und aktuelle Angaben in diesem Werk zum Ausdruck zu bringen, übernehmen weder der Herausgeber, noch der Autor oder andere an der Arbeit beteiligten Personen eine Verantwortung für fehlerhafte Angaben oder deren Folgen. Eventuelle Berichtigungen können erst in der nächsten Auflage berücksichtigt werden.

Bibliographische Informationen der Deutschen Nationalbibliothek
Die Deutsche Nationalbibliothek verzeichnet diese Publikation in der Deutschen Nationalbibliographie; detaillierte bibliographische Daten sind im Internet über *http://dnb.d-nb.de* abrufbar.
1. Aufl. – Göttingen: Cuvillier, 2018

© Cuvillier-Verlag · Göttingen 2018
Nonnenstieg 8, 37075 Göttingen
Telefon: 0551-54724-0
Telefax: 0551-54724-21
www.cuvillier.de

1. Auflage, 2018
Gedruckt auf umweltfreundlichem, säurefreiem Papier aus nachhaltiger Forstwirtschaft.

ISBN 978-3-7369-9881-0
eISBN 978-3-7369-8881-1
ISSN 2199-7756

Riboflavin production with *Ashbya gossypii*: a ^{13}C high-resolution metabolic network analysis under industrial process conditions

Dissertation
zur Erlangung des Grades
der Doktorin der Naturwissenschaften
der Naturwissenschaftlich-Technischen Fakultät
der Universität des Saarlandes

von

Susanne Katharina Schwechheimer

Saarbrücken
2018

Tag des Kolloquiums: 31.10.2018
Dekan: Prof. Dr. Guido Kickelbick
Berichterstatter: Prof. Dr. Christoph Wittmann
                        Prof. Dr. Andriy Luzhetskyy
Vorsitz: Prof. Dr. Gert-Wieland Kohring
Akademischer Mitarbeiter: Dr. Björn Becker

PUBLICATIONS

Partial results of this work have been published in advance authorized by the Institute of Systems Biotechnology (Universität des Saarlandes) represented by Prof. Dr. Christoph Wittmann.

Peer-reviewed articles

Schwechheimer, S. K., Becker, J., Peyriga, L., Portais, J.-C., Wittmann, C., 2018. Metabolic flux analysis in *Ashbya gossypii* using ^{13}C-labeled yeast extract: industrial riboflavin production under complex nutrient conditions. Microb. Cell Fact. 17, 162-184.

Schwechheimer, S. K., Becker, J., Wittmann, C., 2018. Towards better understanding of industrial cell factories: novel approaches for ^{13}C metabolic flux analysis in complex nutrient environments. Curr. Opin. Biotechnol. 54, 128-137.

Schwechheimer, S. K., Becker, J., Peyriga, L., Portais, J.-C., Sauer, D., Müller, R., Hoff, B., Haefner, S., Schröder, H., Zelder, O., Wittmann, C., 2018. Improved riboflavin production with *A. gossypii* from vegetable oil based on ^{13}C metabolic network analysis with combined labeling analysis by GC/MS, LC/MS, 1D, and 2D NMR. Metab. Eng. 47, 357-373.

Schwechheimer, S. K., Park, E. Y., Revuelta, J. L., Becker, J., Wittmann, C., 2016. Biotechnology of riboflavin. Appl. Microbiol. Biotechnol. 100, 2107-2119.

Following **peer-reviewed article** was published during this work, but is not part of this dissertation.

Vassilev, I., Gießelmann, G., **Schwechheimer, S. K.**, Wittmann, C., Virdis, B., Krömer, J. O., 2018. Anodic electro-fermentation: anaerobic production of L-lysine by recombinant *Corynebacterium glutamicum*. Biotechnol. Bioeng. 10, 26562.

Conference contributions

Schwechheimer, S. K., Becker, J., Portais, J.-C., Müller, R., Wittmann, C., Improved riboflavin production with *A. gossypii* from vegetable oil based on ^{13}C metabolic network analysis with combined labeling analysis by GC/MS, LC/MS, 1D, and 2D NMR. March 2018, European Mass Spectrometry Conference, Saarbrücken, Germany.

Schwechheimer, S. K., Becker, J., Peyriga, L., Portais, J.-C., Sauer, D., Müller, R., Hoff, B., Haefner, S., Schröder, H., Zelder, O., Wittmann, C., Improved riboflavin production with *A. gossypii* from vegetable oil based on ^{13}C metabolic network analysis with combined labeling analysis by GC/MS, LC/MS, 1D, and 2D NMR. June 2018, Metabolic Engineering Conference, Munich, Germany.

DANKSAGUNG

Ich möchte mich besonders herzlich bei meinem Doktorvater Prof. Dr. Christoph Wittmann für die Bereitstellung des Themas und die intensive Betreuung der Arbeit bedanken. Danke für die offene Tür, das offene Ohr und die vielen großen und kleinen Dinge, die ich von ihm lernen durfte.

Vielen Dank an Prof. Dr. Andriy Luzhetskyy für die Begutachtung dieser Arbeit. Außerdem möchte ich mich bei Prof. Dr. Gert-Wieland Kohring für den Prüfungsvorsitz und Dr. Björn Becker für die Übernahme des akademischen Beisitzers bedanken.

Bei der BASF SE bedanke ich mich ebenfalls für die Bereitstellung des Themas, die finanzielle Unterstützung und ganz besonders danke ich Dr. Birgit Hoff, Dr. Hartwig Schröder, Dr. Stefan Haefner, Dr. Oskar Zelder und Dr. Marvin Schulz für den regen wissenschaftlichen Austausch. Prof. Jean-Charles Portais und Lindsay Peyriga danke ich für die Durchführung der NMR-Messungen, die den entscheidenden Beitrag zu dieser Arbeit geleistet haben. Prof. Dr. Rolf Müller und Daniel Sauer danke ich für die LC/MS-Messungen von Riboflavin.

Ein großes *merci beaucoup* geht an Michel Fritz für seine Geduld, seine Hilfe bei allen analytischen Problemen und seinen Humor. Ganz herzlich möchte ich mich auch bei Dr. Judith Becker für ihre große Hilfe und die wissenschaftlichen Gespräche bedanken. Ein süßes Dankeschön geht an meine Namensvetterin, Susanne Haßdenteufel, für die wöchentliche Schokoladenauszeit, ihren Einsatz und vor allem alle nicht-wissenschaftlichen Gespräche. Ich bedanke mich ganz herzlich bei allen Kollegen vom iSBio für die tolle Arbeitsatmosphäre. Ein großer Dank geht an meinen ersten und besten Bürokollegen Gideon Gießelmann, dicht gefolgt von Sören Starck und Nadja Barton. Dr. Anna Lange danke ich für all ihr Wissen über die GC/MS und überhaupt. Liebe Jessica, lieber Jonathan, liebe Lisa: Danke für alles!

Für die Unterstützung in wirklich jeder Lebenslage danke ich den besten Freunden, die es gibt, ihr wisst, wer ihr seid. Ich danke meiner Familie. Meiner Mutter bin ich unendlich dankbar für alles, ohne sie wäre ich nie so weit gekommen. Ganz besonders danke ich meiner Schwester Johanna, für ihren emotionalen Rückhalt, ihr Verständnis und den Blick für das Wesentliche. Meiner Freundin Alina danke ich für ihre Geduld, ihre Fähigkeit, mir immer wieder neuen Mut zu geben und ihr großes Herz.

TABLE OF CONTENTS

SUMMARY

The fungus *Ashbya gossypii* is an important industrial producer of riboflavin, i.e. vitamin B_2. Here, we developed and then used a highly sophisticated set-up of parallel ^{13}C tracer studies with labeling analysis by GC/MS, LC/MS, 1D, and 2D NMR to resolve carbon fluxes and obtain a detailed picture of the underlying metabolism in the overproducing strain *A. gossypii* B2 during growth and riboflavin production from a complex industrial medium using vegetable oil as carbon source. Glycine was exclusively used as carbon-two – but not carbon-one (C_1) – donor of the vitamin's pyrimidine ring due to the proven absence of a functional glycine cleavage system. Yeast extract was the main carbon source during growth, while still contributing 8 % overall carbon to riboflavin. Overall carbon flux from rapeseed oil into riboflavin equaled 80 %. Transmembrane formate flux simulations revealed that the C_1-supply displayed a severe bottleneck during the initial riboflavin production, which was overcome in later phases of the cultivation by intrinsic formate accumulation. The transiently limiting C_1-pool was successfully replenished by time-resolved feeding of formate or serine. This increased the intracellular precursor availability and resulted in a riboflavin titer increase of 45 %. This study is the first that successfully sheds light on carbon fluxes during the growth and riboflavin production phase by use of ^{13}C tracers and a complementary platform of analytical techniques.

ZUSAMMENFASSUNG

Der Pilz *Ashbya gossypii* ist ein wichtiger industrieller Produzent für Riboflavin. In dieser Studie wurde eine anspruchsvolle Kombination an parallelen ^{13}C Tracerexperimenten entwickelt und durchgeführt, wobei die Markierungen mittels GC/MS, LC/MS, 1D und 2D NMR analysiert wurden, um die Kohlenstoffflüsse in dem Überproduzenten *A. gossypii* B2 sowohl im Wachstum als auch während der Riboflavinproduktion auf Komplexmedium mit Pflanzenöl als Kohlenstoffquelle, auflösen zu können. Glycin wurde ausschließlich als Zwei-Kohlenstoff-Donor – aber nicht C_1-Donor – für den Pyrimidinring des Vitamins verwendet, was das Fehlen eines funktionalen Glycin-Decarboxylase-Komplexes bewies. Hefeextrakt (YE) war die wichtigste Kohlenstoffquelle während des Wachstums. Der Gesamtkohlenstofffluss von YE und Rüböl zu Riboflavin betrug entsprechend 8 % bzw. 80 %. Simulationen des transmembranen Formiatflusses zeigten, dass die C_1-Bereitstellung während der frühen Riboflavinproduktion limitierend war, was in der späten Kultivierungsphase durch intrinsische Formiatbildung überwunden wurde. Der transient limitierende C_1-Pool wurde erfolgreich durch die zeitaufgelöste Zugabe von Formiat oder Serin aufgefüllt. Dies steigerte den Produkttiter um 45 % durch erhöhte Verfügbarkeit von Vorläufermolekülen. Diese Studie ist die Erste, die Aufschluss über Stoffflüsse während des Wachstums und der Riboflavinproduktion durch den Gebrauch von ^{13}C Tracersubstanzen und komplementären analytischen Techniken gibt.

1 INTRODUCTION

1.1 General Introduction

Riboflavin, also known as vitamin B_2, is a water-soluble compound, which can be synthesized by plants and microorganisms, but is essential for animals as they lack an endogenous biosynthetic pathway. It plays an important role in multiple cellular functions. Pioneering discovery and research dates back almost 150 years, and throughout the past decade, increasing interest has turned riboflavin meanwhile into one of the most important products in biotechnology. The two active forms of riboflavin, flavin adenine dinucleotide (FAD) and flavin mononucleotide (FMN), act as cofactors for oxidoreductases as well as prosthetic groups for enzymes in the β-oxidation pathway (Massey, 2000). Moreover, riboflavin is part of a flavoprotein called cryptochrome, a photoreceptor in charge of the upkeep of the circadian clock (Banerjee and Batschauer, 2005; Miyamoto and Sancar, 1998). Natural sources of riboflavin are for example milk, eggs, and leafy vegetables (O'Neil, 2006). According to the Food and Nutrition Board (1998), the recommended daily allowance of riboflavin lies between 1.1 and 1.3 mg for women and men assuming a healthy person. In humans, riboflavin deficiency amongst other symptoms is associated with skin lesions and corneal vascularization. Riboflavin, which is exclusively synthesized biotechnologically using microorganisms, is mainly used as feed additive (about 70 % of today's market), whereas about 30 % are used as food additive and for pharmaceutical applications (Revuelta et al., 2016). The world market for riboflavin has more than doubled in a little over a decade from 4000 t a^{-1} to 9000 t a^{-1} in 2002 and 2015, respectively (Schwechheimer et al., 2016). Hereby, the production has come a long way from chemical synthesis on fossil fuels to the exclusive biotechnological production today. One of the industrial microbes for riboflavin is *Ashbya gossypii*, a natural overproducer. It accumulates up to 20 g L^{-1} riboflavin, explaining the high interest in this filamentous fungus (Abbas and Sibirny, 2011; Kato and Park, 2012; Lim et al., 2001; Revuelta et al., 2016; Schwechheimer et al., 2016; Stahmann et al., 2000). The chemical company BASF has installed a plant in South Korea, which is specialized in riboflavin production on industrial scale using *A. gossypii* (Schwechheimer et al., 2016). The biochemistry of riboflavin biosynthesis as well as the empirically derived fermentation set-up for *A. gossypii* are extremely complex. Several subcellular compartments contribute to the production of the vitamin, including the peroxisome, the mitochondrion, and the cytosol (Kato and Park, 2012). Furthermore, industrial production is based on a complex mixture of various raw materials: corn steep liquor, peptone, yeast extract in addition to the main carbon source vegetable oil (Epstein et al., 1979; Malzahn et al., 1959; O'Neil, 2006; Tanner et al., 1948). At

1

the starting point of biosynthesis, the major carbon precursors of the vitamin are ribulose 5-phosphate and guanosine triphosphate (GTP). These intermediates are converted into riboflavin in a total number of seven steps. Glucose has been described as alternative carbon source (Demain, 1972; Tanner et al., 1949) and the sugar is the dominant substrate used for research of *A. gossypii* (Bacher et al., 1985; Ledesma-Amaro et al., 2015c; Schlüpen et al., 2003; Silva et al., 2015). So far, classical strain improvement by random mutagenesis and selection (Schmidt et al., 1996a; Schmidt et al., 1996b; Sugimoto et al., 2010) and empirical optimization of the fermentation process (Sahm et al., 2013; Storhas and Metz, 2006) have displayed the dominant strategies towards better production (Abbas and Sibirny, 2011; Kato and Park, 2012; Lim et al., 2001; Revuelta et al., 2016; Schwechheimer et al., 2016; Stahmann et al., 2000). Only a few success cases have managed to improve riboflavin production using rational approaches, such as metabolic engineering (Buey et al., 2015; Jiménez et al., 2008; Monschau et al., 1998; Sugimoto et al., 2009). This might result from the fact that the molecular processes involved in production of the vitamin are only partly understood. Admittedly, a huge complexity is faced inside and outside the cells: a multi-compartment biosynthesis meets a multi-substrate process environment. Without doubt, tailored strain and process engineering, strongly desired to keep up with market demands, will benefit from a better understanding of riboflavin production in *A. gossypii*.

1.2 Objectives

The aim of this work was the resolution of carbon fluxes of growth and riboflavin production by *A. gossypii* under industrial process conditions. Such flux studies typically require well-defined conditions, such as minimal media and only one carbon source. In this regard, the industrial process, i.e. medium with complex compounds and several carbon sources, posed an additional level of complexity. In order to tackle this challenge on a systems biology level, reproducible data are of utmost importance. To this end, a reliable cultivation and sampling scheme should be developed prior to the flux studies. A number of technical developments provided an experimental approach for tailored ^{13}C isotope studies on a complex medium with a deep assessment of ^{13}C labeling in various metabolites by complementary gas chromatography/mass spectrometry (GC/MS), liquid chromatography/mass spectrometry (LC/MS), and nuclear magnetic resonance (NMR) approaches. The integration of the data gained from the various analytical methods should provide deeper insight into the riboflavin metabolism in *A. gossypii* and optimally offer new starting points and targets for process optimization.

2   THEORETICAL BACKGROUND

2.1   Vitamin B₂: discovery and pioneering chemical synthesis

Riboflavin was first mentioned by Blyth in 1879, who isolated a yellow-fluorescing substance from milk whey, which he called lactochrome (Northrop-Clewes and Thurnham, 2012). However, it took almost half a century until the vitamin was isolated, its structure described, and its nutritional function revealed. Early micronutrition studies with rats showed growth impairment upon ariboflavinosis. These findings led to more intensive research in the field of vitamins in general and of riboflavin in particular. In the early 1930s riboflavin was successfully isolated from egg white (Kuhn et al., 1933b), milk whey, and vegetables (Eggersdorfer et al., 2012; Kuhn et al., 1933a), which was followed by the discovery of its structure (then called lactoflavin): a methylated isoalloxazine ring with a ribityl-sidechain (Figure 1) (Karrer et al., 1935; Kuhn, 1936). Soon afterwards, vitamin B₂ officially became "riboflavin" (from Latin *flavus* for yellow and "ribo" for the ribityl-sidechain) by the Council of Pharmacy and Chemistry of the American Medical Association (Northrop-Clewes and Thurnham, 2012). The high interest in the molecule was a major driver to derive it by chemical synthesis (Karrer et al., 1935; Kuhn, 1936). The initially developed and still major multi-step large-scale chemical route starts either from D-glucose or D-ribose (Figure 1). Glucose is first oxidized to arabonate, which is subsequently epimerized to ribonate and transformed into ribonolactone. This intermediate is reduced to D-ribose using amalgam. After addition of xylidine, the product ribitylxylidine and an aniline derivative together form phenylazo-ribitylxylidine. The final reaction step of the chemical synthesis is a cyclocondensation of phenylazo-ribitylxylidine with barbituric acid, which yields riboflavin as a product (Wolf et al., 1983). For many years, with a few alterations, this was the only way to synthesize riboflavin. Nowadays, the production of riboflavin is exclusively done using fermentation as it is economically and ecologically more feasible. While the production of riboflavin via fermentation was only 5 % of the annual production in 1990, the percentage of biotechnological production has increased to 75 % of the market volume in 2002 within just twelve years due to metabolic engineering of the production strains (Schwechheimer et al., 2016). The two dominating processes employ the Gram positive bacterium *Bacillus subtilis* and the hemiascomycete *Ashbya gossypii*. Today, due to further optimization of the bio-based production in the past years, chemical synthesis has been replaced completely. Most of today's riboflavin market is used as feed additive (Sahm et al., 2013), but also as food fortification, dietary supplement, pharmaceutical applications as well as food colorant (E-101) in yoghurt and drinks.

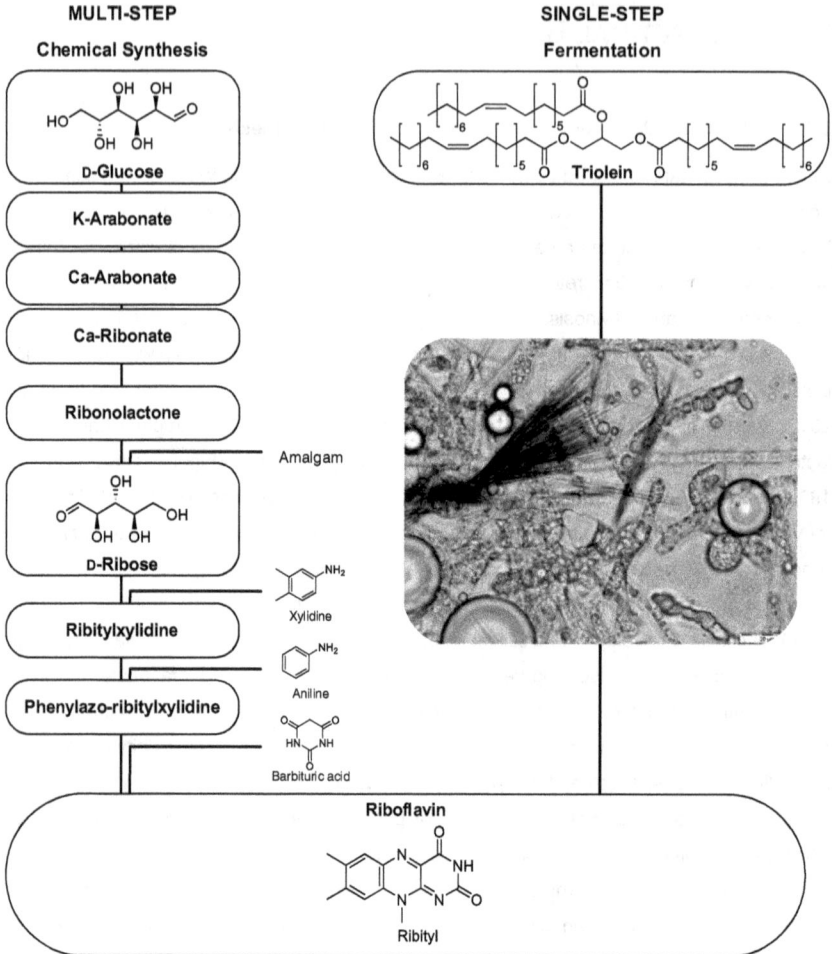

MULTI-STEP
Chemical Synthesis

SINGLE-STEP
Fermentation

Figure 1: Chemical versus biotechnological riboflavin synthesis by fermentation. The microscopic image depicts a wild type *A. gossypii* in a vegetable oil medium. Mycelia, oil droplets, and riboflavin crystals are shown in a 600-fold magnification (Schwechheimer et al., 2016).

2.2 *Ashbya gossypii* – a fungal riboflavin overproducer

The filamentous hemiascomycete *A. gossypii* was first isolated from cotton as phytopathogen causing stigmatomycosis in 1926 in the British West Indies (Ashby and Nowell, 1926). Later, it was also discovered on other crops in tropic and sub-tropic regions (Pridham and Raper,

1950). The use of insecticides efficiently reduced the role of *A. gossypii* as phytopathogen, because the fungus relies on insect vectors for transmission. It is incapable of penetrating intact plant cell walls, which can be explained by the low amount of extracellular enzymes secreted by *A. gossypii* (Aguiar et al., 2014b). Its ability to produce large amounts of volatile aroma compounds might be advantageous when attracting insects for transmission (Ravasio et al., 2014; Wendland et al., 2011).

The life cycle of *A. gossypii* starts from a needle-shaped spore. After germ bubble and germ tube formation, a mycelium is generated by lateral branching of the initial hyphae (Figure 2A), which also leads to dichotomous tip branching. Each hyphal cell, which is separated by septa from neighboring cells, contains several haploid nuclei. In the late growth phase, sporangia usually containing eight spores each are formed by fragmentation of old hyphae. These spores, which are linked by actin filaments, are then released (Figure 2C) (Wendland and Walther, 2005). The transition from vegetative growth to sporulation is linked to nutrient depletion, however, the exact underlying mechanisms are still poorly understood. Recent studies on the genome suggest that *A. gossypii* displays two different mating types and thus, may also exhibit sexual reproduction (Dietrich et al., 2013).

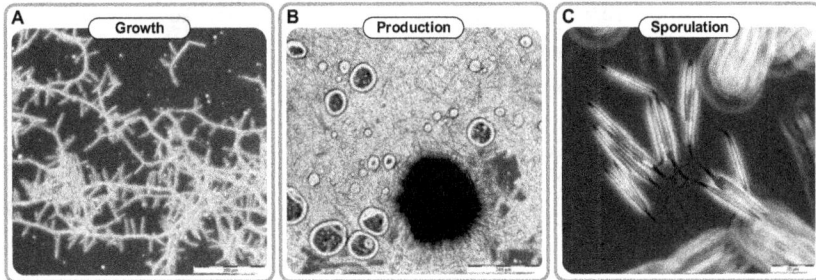

Figure 2: Microscopic images of *A. gossypii* B2 grown on complex medium and vegetable oil (A, B) or glucose (C). 6 h after inoculation, lateral branching can be observed in the growing culture (A). The low solubility of riboflavin leads to formation of vitamin crystals after 72 h (B). At the end of cultivation on glucose, the needle-shaped spores, which are connected by actin, are released from sporangia (C). Images A and B are shown in a 100-fold magnification. The spores in image C are 600-fold magnified.

A. gossypii belongs to the family of Saccharomycetaceae and shares more than 90 % synteny of protein coding genes with the budding yeast *Saccharomyces cerevisiae*. Due to its very small genome size (9 Mb), close relationship with *S. cerevisiae*, and readily available molecular tools, it has become a model organism for filamentous growth (Wendland and Walther, 2005).

Overproduction of riboflavin by *A. gossypii* was first reported in 1946. The production starts once vegetative growth has ceased and cells start sporulating (Figure 2B). Originally, it was hypothesized that overproduction of riboflavin protected the hyaline spores against ultraviolet (UV) light (Stahmann et al., 2001). However, since also non-sporulating cells accumulate riboflavin, these two events are not exclusively connected (Nieland and Stahmann, 2013; Walther and Wendland, 2012). Over the past years, it was shown that nutritional and oxidative stress induce riboflavin overproduction in the fungus (Kavitha and Chandra, 2014; Schlösser et al., 2007). Walther and Wendland (2012) even suggested that riboflavin plays a role in stress defence against oxidative burst of plants upon infection with *A. gossypii* cells.

In recent years, the potential of *A. gossypii* as host for other high-value products has been investigated and its spectrum was tremendously broadened (Aguiar et al., 2017). This entails flavor compounds derived from the Ehrlich pathway (2-phenylethanol and isoamyl alcohol) (Ravasio et al., 2014) and purine biosynthesis (inosine and guanosine contribute to the *umami* flavor) (Ledesma-Amaro et al., 2015a; Ledesma-Amaro et al., 2016), but also single cell oil (SCO) production (Díaz-Fernández et al., 2017; Ledesma-Amaro et al., 2015b; Ledesma-Amaro et al., 2014b; Lozano-Martínez et al., 2016). The expression performance of recombinant cell wall degrading enzymes was inspected (Aguiar et al., 2014a; Ribeiro et al., 2010). In contrast to other filamentous fungi, *A. gossypii* secretes only a small amount of extracellular enzymes, which makes it a promising host for recombinant protein production. However, current titers still leave room for improvement. Most recently, the overproduction of folic acid (vitamin B_9) was reported (Serrano-Amatriain et al., 2016). Genetic engineering towards vitamin B_9 production seems obvious, considering the high potential of the purine biosynthetic pathway in this organism and the shared precursor, GTP, with riboflavin.

2.3 Riboflavin biosynthesis – pathways and regulations

Many microorganisms are capable to synthesize riboflavin. *Candida famata*, *Clostridium acetobutylicum*, and *Lactobacillus fermentum* are able to overproduce the vitamin (Demain, 1972). It is, however, easy to understand that most of the knowledge on riboflavin biosynthesis has been collected for the two major industrial producers: *A. gossypii* (Figure 5) and *B. subtilis* share two important precursors for riboflavin biosynthesis: ribulose 5-phosphate (Ru5P) is derived from the pentose phosphate (PP) pathway and GTP originates in the purine biosynthesis (Bacher et al., 2000).

2.3.1 Terminal biosynthesis

The terminal riboflavin biosynthetic chain comprises a total of seven enzymatic steps starting from two different branches: the purine biosynthesis and the PP pathway (Figure 3). The GTP cyclohydrolase catalyzes the cleavage of GTP with release of formate. This step is encoded by *RIB1* in *A. gossypii* and by *ribA* in *B. subtilis*. The first reaction is followed by a reduction reaction in the fungus, carried out by the gene product of *RIB7* (DARPP reductase), and a subsequent deamination (*RIB2*, DarPP deaminase). In *B. subtilis*, these latter two steps are in reverse order and are catalyzed by a bifunctional enzyme encoded by *ribG*. The phosphatase that cleaves ArPP (2,5-diamino-6-ribityl-amino-2,4(1H,3H)pyrimidinedione 5'-phosphate) into ArP (5-amino-6-ribitylamino-2,4(1H,3H)-pyrimidinedione) is the only, still unknown, enzyme in the riboflavin biosynthetic pathway. Ribulose 5-phosphate is converted to DHBP (3,4-dihydroxy-2-butanone 4-phosphate) by the DHBP synthase (*RIB3* or *ribA* for *A. gossypii* and *B. subtilis*, respectively). At this point the two different branches of the riboflavin pathway merge into one. The condensation of DHBP and ArP yields one molecule of DRL (6,7-dimethyl-8-ribityllumazine) and is catalyzed by the lumazine synthase (*RIB4* for *A. gossypii*, *ribH* for *B. subtilis*). In the final step of the riboflavin biosynthetic pathway the enzyme riboflavin synthase converts two mole DRL into one mole riboflavin and one mole ArP, which is recycled in the previous step (Fischer and Bacher, 2005).

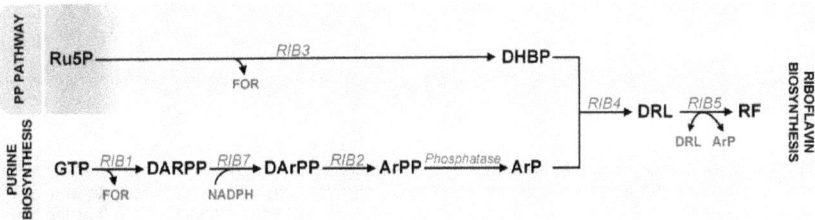

Figure 3: Terminal biosynthesis of riboflavin in *A. gossypii*. Ru5P is converted into DHBP and condensed with ArP to yield DRL. In the final step, riboflavin synthase catalyzes the conversion of DRL to riboflavin. Note that the pentose phosphate pathway branch is undergone twice, since two molecules DHBP are needed for one molecule of riboflavin. ArP, 5-amino-6-ribitylamino-2,4(1H,3H)-pyrimidinedione; ArPP, 5-amino-6-ribitylamino-2,4(1H,3H)-pyrimidinedione 5-phosphate; DARPP, 2,5-diamino-6-ribosylamino-4(3H)-pyrimidinone 5-phosphate; DarPP, 2,5-diamino-6-ribitylamino-pyrimidinone 5-phosphate; DHBP, 3,4-dihydroxybutanone 4-phosphate; DRL, 6,7-dimethyl-8-ribityllumazine; FOR, formate; GTP, guanosine triphosphate; NADPH, nicotinamide adenine dinucleotide phosphate; PP pathway, pentose phosphate pathway; Ru5P, ribulose 5-phosphate; RF, riboflavin; *RIB1*, GTP cyclohydrolase II; *RIB2*, DarPP deaminase; *RIB3*, DHBP synthase; *RIB4*, lumazine synthase; *RIB5*, riboflavin synthase; *RIB7*, DARPP reductase.

7

2.3.2 Precursor supply

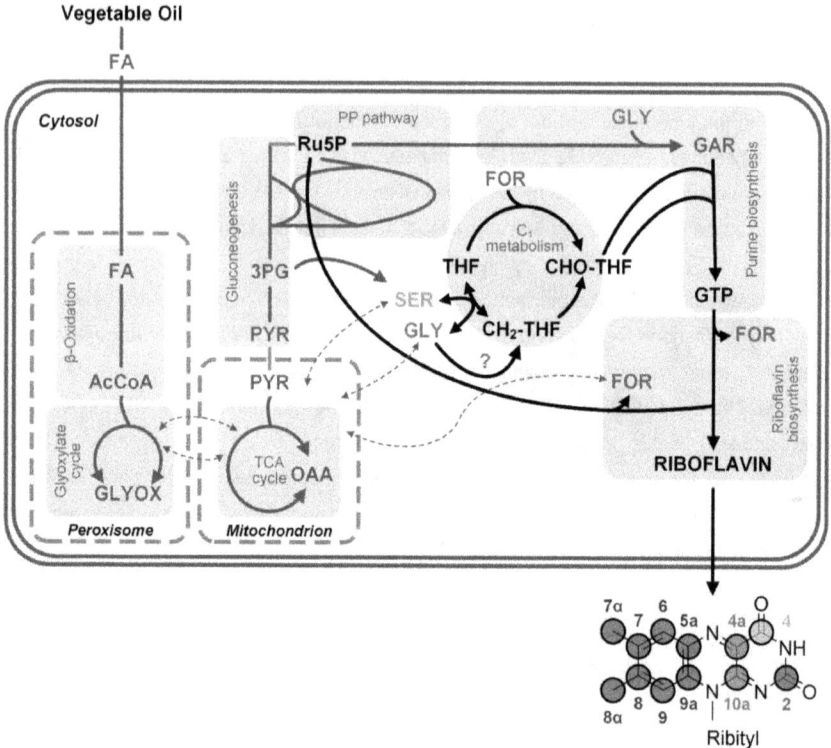

Figure 4: Schematic riboflavin biosynthesis from vegetable oil in *A. gossypii*. The multi-compartment process ends with the terminal riboflavin biosynthesis starting from ribulose 5-phosphate and GTP. Dashed lines indicate exchange of metabolites between different compartments. Note that the one-carbon metabolism is only drawn in the cytosol. It can be assumed, however, that there is also a one-carbon metabolism in the mitochondrion. The circles in the riboflavin molecule designate the carbon atoms and their respective molecular origin: ribulose 5-phosphate (dark grey), carbon dioxide (light grey), glycine metabolism (blue), one-carbon metabolism (red). The ribityl side chain originates from ribose 5-phosphate. 3PG, 3-phosphoplycerate; CH$_2$-THF, 5,10-methylenetetrahydrofolate; AcCoA, acetyl-CoA; CHO-THF, 10-formyltetrahydrofolate; FA, fatty acid; FOR, formate; GAR, glycineamide ribonucleotide; GLY, glycine; GTP, guanosine-5'-triphosphate; OAA, oxaloacetate; PYR, pyruvate; Ru5P, ribulose 5-phosphate; SER, serine; THF, tetrahydrofolate.

For the overproduction of riboflavin, *A. gossypii* prefers oil as substrate, which is cleaved into fatty acids and glycerol by an extracellular lipase (Stahmann et al., 1997). The fatty acids are transported into the cell and oxidized into acetyl-CoA via the β-oxidation pathway, located in the peroxisome. Acetyl-CoA is further metabolized via the glyoxylate shunt, gluconeogenesis, and the PP pathway (Figure 4, Figure 5). In addition, supplementation with glycine is frequently used for riboflavin production (Demain, 1972; Malzahn et al., 1959). The amino acid is supposed to have two functions. One molecule glycine is incorporated into the pyrimidine ring of riboflavin (Plaut, 1954a). Furthermore, glycine is linked to the folate-dependent carbon-one (C_1) metabolism, which supplies 10-formyltetrahydrofolate (CHO-THF) as essential C_1 donor for the biosynthesis (Pasternack et al., 1996; Schlüpen et al., 2003). In more detail, the enzyme serine hydroxymethyltransferase (SHMT) forms serine from glycine. The terminal carbon atom of serine then contributes to the C_1 pool via 5,10-methylenetetrahydrofolate (CH_2-THF), which in turn is converted into CHO-THF, the immediate C_1 donor for riboflavin (Figure 4). Moreover, a potential glycine cleavage system (GCS) has been suggested as alternative route in the fungus (Schlüpen et al., 2003), but to date has neither been proven nor disproven experimentally.

2.3.3 Regulation

Regulation of the riboflavin biosynthetic pathway is not completely solved for the hemiascomycete *A. gossypii*. However, a few studies have dealt with unraveling the regulatory mechanisms behind riboflavin overproduction, which has been linked to nutritional (Schlösser et al., 2007) as well as oxidative stress (Kavitha and Chandra, 2009; Kavitha and Chandra, 2014). It was reported that the beginning of riboflavin oversynthesis and sporulation were linked, which was proven by addition of the second messenger cyclic adenosine monophosphate (cAMP) that is known to inhibit sporulation in fungi: riboflavin biosynthesis as well as sporulation were negatively affected (Stahmann et al., 2001). Supplementation of riboflavin to spore suspensions had a positive effect on spore viability upon UV light exposure (Stahmann et al., 2001). The regulation of pathway-specific enzymes was also investigated. DHBP synthase (encoded by the gene *RIB3*) catalyzes the first step in riboflavin biosynthesis starting from the ribulose 5-phosphate derived branch and carries twice the metabolic burden compared to the GTP branch of this pathway. Therefore, its regulation was of special interest. *RIB3* exhibited increased mRNA levels during the riboflavin production phase, caused by induction of the promoter (Schlösser et al., 2001). In a different study, upregulation of the three *RIB* genes involved in the PP pathway branch (*RIB3, RIB4, RIB5*) were upregulated upon cessation of growth due to nutrient depletion and entry into the riboflavin production phase (Schlösser et al., 2007). A more recent study, however, reported that there is no significant increase at the transcriptional level for all *RIB* genes except *RIB4* during the riboflavin biosynthetic phase (Ledesma-Amaro et al., 2015c). Two transcription factors have been described in the literature as having direct or indirect regulatory functions in the production of riboflavin: *BAS1* and *YAP1*. The Myb-related factor *BAS1* is involved in the adenine-dependent transcriptional control of the genes *SHM2* and *ADE4*, which play important roles in the glycine and purine metabolism, respectively, both of which contribute to the riboflavin precursor GTP.

A C-terminal deletion of the *BAS1* gene resulted in increased riboflavin titers due to constitutive activation of *SHM2* and *ADE4* (Mateos et al., 2006). The second transcription factor, *YAP1*, is known to be involved in oxidative stress response and targeted exposure of *A. gossypii* to oxidative stress led to increased riboflavin titers in a Yap1-dependent manner. The Yap-regulon comprises more than 100 genes, amongst others *RIB4*. This gene of the riboflavin pathway contains three Yap-binding domains and is transcriptionally controlled by *YAP1* (Walther and Wendland, 2012).

2.3.4 Biosynthesis in *B. subtilis*

Since *B. subtilis* is the other major industrial riboflavin producer next to *A. gossypii*, its riboflavin biosynthesis should be discussed briefly. Starting from its preferred carbon source glucose *B. subtilis* forms glucose 6-phosphate, which then enters into the PP pathway toward riboflavin biosynthesis. The bacterial riboflavin biosynthesis is organized in the so-called *rib* operon, which entails five genes (*ribGBAHT*) (Abbas and Sibirny, 2011; Perkins et al., 1999; Yakimov et al., 2014). The riboflavin biosynthesis in *B. subtilis* includes two bifunctional enzymes: the gene product of *ribA* shows GTP cyclohydrolase II activity as well as DHBP synthase activity (Hümbelin et al., 1999). The second bifunctional enzyme is the above mentioned gene product of *ribG*: the combined reductase and deaminase (Richter et al., 1997). As for *A. gossypii*, the regulation of riboflavin biosynthesis is not fully understood in *B. subtilis*, either. The *rib* operon of *B. subtilis* itself at least seems to be regulated by a "riboswitch" (Mironov et al., 2002). A conserved sequence within the 5'-untranslated region of the *rib* operon is likely to fold into a secondary structure. FMN, which is the product of a kinase reaction of the gene product of *ribC*, is able to directly bind to this secondary structure, thus repressing transcription of the *rib* operon (Mack et al., 1998; Mironov et al., 2002). RibR, a protein that is not part of the *rib* operon, is believed to act as a regulatory protein since it seems to be able to bind to this riboswitch (Higashitsuji et al., 2007). The gene *ribR* is part of a transcription unit that entails gene products involved in sulphur uptake and degradation. Recently it was shown that when sulphur is present, *ribR* expression increases, the FMN demand of the cell rises, and the *rib* operon is expressed even with high FMN levels (Pedrolli et al., 2015).

2.4 Biotechnology and industrial production of riboflavin

2.4.1 Metabolic engineering of *A. gossypii*

The availability and advance in the development of molecular tools have made it very easy to modify and engineer prokaryotes and eukaryotes like *B. subtilis* and the model organism for filamentous fungi *A. gossypii* (Figure 5, Table 1). In *A. gossypii*, riboflavin production was increased in the wild type by medium supplementation with riboflavin precursors such as glycine or hypoxanthine (Stahmann et al., 2000). Since structural analogs of purines or riboflavin were not suitable for the selection of overproducing strains, other anti-metabolites were successfully implemented. Itaconate and oxalate, both inhibitors of the key-enzyme of the glyoxylic pathway isocitrate lyase (ICL), were applied to yield higher producing strains. (Schmidt et al., 1996a; Schmidt et al., 1996b; Sugimoto et al., 2010). Early metabolic engineering strategies included the overexpression of GLY1. This gene codes for a threonine aldolase and catalyzes the formation of glycine from threonine. An additional supplementation of threonine had a positive effect on riboflavin biosynthesis as well (Monschau et al., 1998). A. gossypii tends to accumulate riboflavin intracellularly in vacuoles, which is counterproductive especially for industrial processes. Therefore, the vacuolar ATPase subunit A (VMA1) was disrupted, leading to an increase in secreted riboflavin. Interestingly, disruption of the same gene in S. cerevisiae is lethal (Förster et al., 1999). Research interest in metabolic engineering of A. gossypii has focused on the PP pathway, the glycine biosynthetic pathway, and the purine biosynthesis. Overexpression of the PRS gene (PRPP synthetase) as well as ADE4 (PRPP amidotransferase) increased the carbon flux through the PP and purine biosynthetic pathways (Jiménez et al., 2005; Jiménez et al., 2008). Knock-out of a Myb-related transcription factor also led to an enhanced carbon flow through the purine biosynthesis to GTP as immediate precursor of riboflavin as well as glycine biosynthesis (Mateos et al., 2006). A successful strategy to increase glycine precursor supply has been the deletion of SHM2, a gene that codes for a serine hydroxymethyltransferase in A. gossypii. The deletion mutant showed a reduced activity for the conversion of glycine to serine and ensured a larger glycine pool and 10-fold increase in the riboflavin titer (Schlüpen et al., 2003). Since A. gossypii uses vegetable oil as substrate for riboflavin synthesis, the glyoxylic pathway plays a crucial role in its metabolism. Overexpression of the malate synthase MLS1 improved riboflavin productivity about 1.7-fold (Sugimoto et al., 2009). Recently, the alteration of the pyrimidine pathway resulted in an increased precursor supply for the purine biosynthesis and thus riboflavin production (Silva et al., 2015). Disparity mutagenesis rendered a strain with higher levels of the gene products of ADE1, RIB1, and RIB5, which also led to more riboflavin (Park et al., 2011). Recently, all RIB genes but RIB4 have been overexpressed in A. gossypii, leading to a 3.1-fold increase in riboflavin compared to the wild type. The riboflavin titer could be enhanced even more by underexpression of ADE12, a gene, which codes for an enzyme that catalyzes the conversion of inosine monophosphate (IMP) to adenosine monophosphate (AMP). Together, these modifications could increase the final riboflavin concentration by a factor of 5.4 (Ledesma-Amaro et al., 2015c). In addition, GTP precursor supply and thus riboflavin synthesis could be enhanced by overexpression of the IMP dehydrogenase, which plays a key role in the nucleotide biosynthesis (Buey et al., 2015). Overall this has resulted in strains, which accumulate up to 700 mg L^{-1} of riboflavin (Table 1).

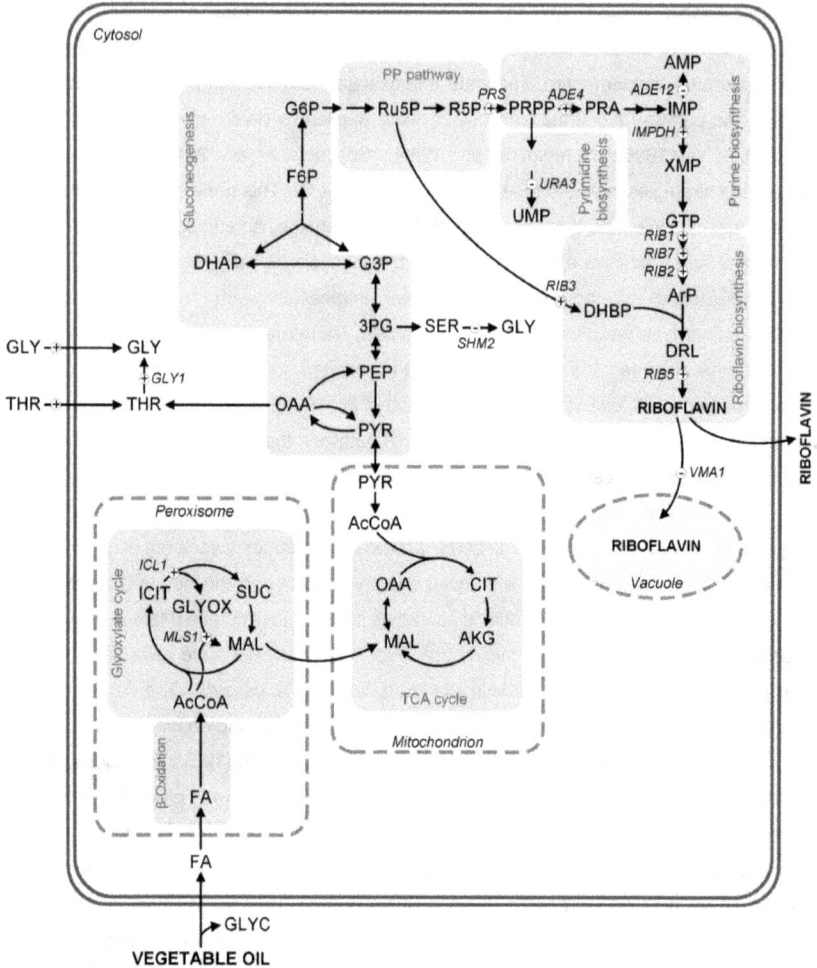

Figure 5: Riboflavin biosynthesis in *Ashbya gossypii*. The symbols + and – indicate increased or decreased fluxes or activities both with positive effect on riboflavin biosynthesis. Note that the conversion of citrate to isocitrate via aconitase most likely does not occur in the peroxisome (Murakami and Yoshino, 1997). 3PG, 3-phosphoglycerate; AcCoA, acetyl-coenzyme A; AKG, α-ketoglutarate; AMP, adenosine monophosphate; ArP, 5-Amino-6-ribitylamino-2,4(1H,3H)-pyrimidinedione; CIT, citrate; DHAP, dihydroxyacetone phosphate; DHBP, 3,4-Dihydroxy-2-butanone-4-phosphate; DLR, 6,7-Dimethyl-8-ribityllumazine; F6P, fructose 6-phosphate; FA, fatty acids; G3P, glyceraldehyde 3-phosphate; G6P, glucose 6-phosphate; GLY, glycine; GLYC, glycerol; GLYOX, glyoxylate; GTP, guanosine triphosphate; ICL1, isocitrate lyase; IMP, inosine monophosphate; MAL, malate; OAA, oxaloacetate; OMP, orothidine monophosphate; PEP, phosphoenolpyruvate; PP pathway, pentose phosphate pathway; PRA, 5-phosphpribosylamine; PRPP, phosphoribosylpyrophosphate; PYR, pyruvate; R5P, ribose 5-phosphate; Ru5P, ribulose 5-phosphate; SER, serine; TCA cycle, tricarboxylic acid cycle; THR, threonine; UMP, uridine monophosphate; XMP, xanthosine monophosphate; *ADE4*, PRPP amidotransferase; *ADE12*, adenylosuccinate synthase; *GLY1*, threonine aldolase; *ICL1*, isocitrate lyase; *IMPDH*, IMP dehydrogenase; *MLS1*, malate synthase; *PRS*, PRPP synthetase; *RIB1*, GTP cyclohydrolase II; *RIB2*, DarPP deaminase; *RIB3*, DHBP synthase; *RIB5*, riboflavin synthase; *RIB7*, DARPP reductase; *SHM2*, serine hydroxymethyltransferase; *URA3*, orotidine-5-phosphate decarboxylase; *VMA1*, vacuolar ATPase subunit A.

Table 1: Metabolic engineering of *Ashbya gossypii* for riboflavin overproduction by gene manipulation. Except for the overexpression of malate synthase (rapeseed oil), all engineered strains were tested on medium with glucose as main carbon source.

Target gene	Modification	Performance	Reference
GLY	Overexpression	33 mg L^{-1} (3.7 mg L^{-1})[b]	(Monschau et al., 1998)
SHM2	Disruption	96 mg L^{-1} (9 mg L^{-1})[b]	(Schlüpen et al., 2003)
ADE4	Overexpression	77.2 mg L^{-1}/228 mg L^{-1} [a] (28 mg L^{-1})[b]	(Jiménez et al., 2005)
PRS2,4	Overexpression	42.4 mg L^{-1}/48.6 mg L^{-1} [a] (28 mg L^{-1})[b]	(Jiménez et al., 2008)
PRS3	Overexpression	40.4 mg L^{-1}/51.6 mg L^{-1} [a] (28 mg L^{-1})[b]	(Jiménez et al., 2008)
MLS1	Overexpression	700 mg L^{-1} (400 mg L^{-1})[b]	(Sugimoto et al., 2009)
URA3	Disruption	7.5 mg g$_{mycel}$$^{-1}$ (1.0 mg g$_{mycel}$$^{-1}$)[b]	(Silva et al., 2015)
RIB1,2,3,5,7	Overexpression	327 mg L^{-1} (105 mg L^{-1})[b]	(Ledesma-Amaro et al., 2015c)
ADE12	Underexpression	260 mg L^{-1} (98 mg L^{-1})[b]	(Ledesma-Amaro et al., 2015c)
IMPDH	Overexpression	18.9 mg g$_{CDW}$$^{-1}$ (13.8 mg g$_{CDW}$$^{-1}$)[b]	(Buey et al., 2015)

CDW, cell dry weight; mycel, mycelium; *ADE4*, PRPP amidotransferase; *ADE12*, adenylosuccinate synthase; *GLY1*, threonine aldolase; *ICL1*, isocitrate lyase; *IMPDH*, inosine monophosphate dehydrogenase; *MLS1*, malate synthase; *PRS*, PRPP synthetase; *RIB1*, GTP cyclohydrolase II; *RIB2*, DarPP deaminase; *RIB3*, DHBP synthase; *RIB5*, riboflavin synthase; *RIB7*, DARPP reductase; *SHM2*, serine hydroxymethyltransferase; *URA3*, orotidine-5 – phosphate decarboxylase.

[a] Overexpression plus point mutation
[b] Wild type

2.4.2 Metabolic engineering of other microorganisms

B. subtilis, which is not a natural overproducer of riboflavin, was at first optimized using classical breeding like exposure to purine or riboflavin analogs. Mutants resistant to the purine analog 8-azaguanine were able to produce more riboflavin due to deregulation of purine synthesis. *B. subtilis* mutants with a resistance to methionine sulfoxide as well as decoyinine showed an increased flux from IMP to GMP, probably caused by upregulation of the respective synthetase (Matsui et al., 1977; Matsui et al., 1979). The use of roseoflavin as structural analog to riboflavin rendered overproducing mutants that had a deregulated *rib* operon due to mutations in *ribC* or *ribO*. *RibC* is a *trans*-acting regulator of the operon whereas *ribO*, the leader of the 5'-untranslated region of the *rib* operon, functions in a *cis* manner to regulate the operon (Coquard et al., 1997; Mack et al., 1998; Stahmann et al., 2000). Genetic engineering for a commercially competitive *B. subtilis* strain was achieved through *rib* operon duplication as well as substitution of the existing promoter with a phage-derived constitutive promoter (Table 2) (Perkins et al., 1999).

Table 2: Metabolic engineering of *Bacillus subtilis*, *Escherichia coli*, and *Corynebacterium glutamicum* for riboflavin production by gene manipulation. All engineered strains were tested on glucose for overproduction.

Target gene	Modification	Performance	Reference
B. subtilis			
rib operon	Multiple copies, phage promoter	14 g L^{-1} (0.02-0.05 g L^{-1})[a]	(Perkins et al., 1999)
ribA	Overexpression	17.5 g L^{-1}	(Hümbelin et al., 1999)
cyd	Deletion	12.3 g L^{-1} (8.9 g L^{-1})[b]	(Zamboni et al., 2003)
gdh	Overexpression	0.047 g g_{CDW}^{-1} (0.03 g g_{CDW}^{-1})[b]	(Zhu et al., 2006)
ccpn	Deletion	0.062 g g_{Glc}^{-1} (0.038 g g_{Glc}^{-1})[b]	(Tännler et al., 2008b)
purFMNHD	Overexpression	0.031 g g_{Glc}^{-1} (0.025 g g_{Glc}^{-1})[b]	(Shi et al., 2009b)
prs, ywlF	Overexpression	15.0 g L^{-1} (12.0 g L^{-1})[b]	(Shi et al., 2009a)
zwf	Overexpression	0.05 g g_{Glc}^{-1} (0.04 g g_{Glc}^{-1})	(Duan et al., 2010)
purR	Deletion	826 mg L^{-1} (275 mg L^{-1})[a]	(Shi et al., 2014)
gapB, fbp	Overexpression	13.4 g L^{-1} (10.5 g L^{-1})	(Wang et al., 2014)
E. coli			
zwf[1], *gnd*[1], *pgl*[3], *pgi*[2], *edd*[2], *eda*[2], *acs*[3], *ribF*[4]	1) Heterologous expression, 2) deletion, 3) overexpression, 4) underexpression	1036 mg L^{-1} (229 mg L^{-1})	(Lin et al., 2014)
C. glutamicum			
sigH	Overexpression	32.4 µmol L^{-1} (< 5 µmol L^{-1})	(Taniguchi and Wendisch, 2015)

Glc, glucose; *acs*, acetyl-CoA synthetase; *ccpn*, transcriptional regulator of *gapB* and *pckA*; *cyd*, cytochrome *bd* oxidase; *edd*, phosphogluconate dehydratase; *eda*, multifunctional 2-keto-3-deoxygluconate 6-phosphate aldolase and 2-keto-4-hydroxyglutarate aldolase and oxaloacetate decarboxylase; *fbp*, fructose-1,6-bisphosphatase; *gapB*, glyceraldehyde-3-phosphate dehydrogenase; *gdh*, glucose dehydrogenase; *gnd*, 6-phosphogluconate dehydrogenase; *guaA*, GMP synthase; *guaB*, inosine-monophosphate dehydrogenase; *pgi*, glucose-6-phosphate isomerase; *pgl*, 6-phosphogluconolactonase; *prs*, PRPP synthetase; *purA*, adenylosuccinate synthetase; *purD*, phosphoribosylglycinamide synthetase; *purF*, glutamine phosphoribosylpyrophosphate amidotransferase; *purH*, phosphoribosylaminoimidazole carboxy formylformyltransferase/inosine-monophosphate cyclohydrolase; *purM*, phosphoribosylaminoimidazole synthetase; *purN*, phosphoribosylglycinamide formyltransferase; *purR*, transcriptional repressor of *pur* operon; *rib* operon, riboflavin operon; *ribA*, GTP cyclohydrolase II and DHBP synthase; *ribF*, bifunctional riboflavin kinase/FMN adenylyltransferase; *sigH*, sigma factor H; *ywlF*, ribose-5-phosphate isomerase; *zwf*, glucose 6-phosphate dehydrogenase.

[a] Wild type
[b] Parental producer strain

In addition to those modifications, the overexpression of *ribA*, which encodes the rate-limiting steps of GTP cyclohydrolase II and DHBP synthase, led to a significant increase in riboflavin titer of about 1.25-fold (Hümbelin et al., 1999). In the following years, research interest shifted away from the terminal riboflavin synthesis and towards increasing energy and precursor supply. The wild type of *B. subtilis* consumes a lot of energy for its maintenance. By a knockout of a terminal oxidase, this feature could be decreased and energy supply increased (Zamboni et al., 2003). Overexpression of following the enzymes led to a substantial increase of carbon flow through the PP pathway: glucose dehydrogenase (Zhu et al., 2006), glucose-6-phosphate dehydrogenase (Duan et al., 2010), ribose-5-phosphate isomerase and phosphoribosylpyrophosphate (PRPP) synthetase (Shi et al., 2009a). Flux through the purine biosynthesis, with GTP as immediate riboflavin precursor, was genetically engineered by co-overexpression of five genes in the *pur* operon: *purF*, *purM*, *purN*, *purH*, and *purD* (Table 2), which code for enzymes that catalyze reactions using cosubstrates like glutamine or glycine (Shi et al., 2009b). Furthermore, carbon flux through the purine biosynthetic pathway and riboflavin titer could be increased by disrupting the repressor of the *pur* operon (*purR*) (Shi et al., 2014). Concerning product export, often identified as a bottleneck in overproducing strains (Kind et al., 2011), a riboflavin transporter from *Streptomyces davawensis*, which was codon optimized for *B. subtilis*, was successfully expressed in *B. subtilis* and catalyzed riboflavin excretion into the culture medium (Hemberger et al., 2011). This might be an interesting starting point for further strain optimizations. Recently, genetically engineered riboflavin-overproducing *B. subtilis* was successfully applied in a bacterial consortium with *Escherichia coli* and *Shewanella oneidensis*, generating electricity from glucose (Liu et al., 2017). Riboflavin, in that case, functioned as electron shuttle, which illustrates nicely the potential of riboflavin overproduction taken to the next level of bioeconomy.

Another recent example of successful metabolic engineering is *Escherichia coli*, which does not accumulate riboflavin under natural conditions. A basic riboflavin producer strain (RF01S, with the *rib* operon under the control of an inducible promoter) was further genetically engineered to produce larger amounts of vitamin B$_2$. Following modifications were implemented: expression of *Corynebacterium glutamicum zwf* (glucose-6-phosphate dehydrogenase) and *gnd* (6-phosphogluconate dehydrogenase), deletion of *pgi* (glucose-6-phosphate isomerase), and genes of Entner-Doudoroff-pathway, overexpression of *acs* (acetyl-CoA-synthetase), and modulation of expression of *ribF*, a riboflavin kinase (Table 2). After optimizing fermentation conditions a final riboflavin concentration of 2700 mg/L was reached (Lin et al., 2014). More recently, Taniguchi and Wendisch (2015) investigated the relationship between sigma factor H and riboflavin biosynthesis in the soil bacterium *C. glutamicum*. Overexpression of the sigma factor resulted in slight accumulation of riboflavin (Table 2), an interesting proof-of-concept which might be exploited further.

2.4.3 Bioprocess engineering and industrial production of riboflavin

From the late 1950s, the pharmaceutical company Merck established a bio-based process with *Eremothecium ashbyii*, later the related strain *A. gossypii* (Malzahn et al., 1959). However, large-scale riboflavin production using fermentation was not conducted until 1990, when the German company BASF commercialized the *A. gossypii*-process. For the following years, chemical and biotechnological syntheses were performed in parallel, when the chemical plant was eventually shut down. In 2003, the production of riboflavin using *A. gossypii* was moved from Germany to a new production site in Gunsan, Korea. The industrial process is carried out at around 30 °C in 100 m^3 fed-batch fermenters using vegetable oils as carbon source and complex nitrogen sources like soy flour or corn steep liquor and lasts about eight days (Figure 6) (Malzahn et al., 1959; Tanner et al., 1948). The process is divided into two phases: a growth and subsequent production phase. The feed medium ensures supplementation with limited precursors like glycine (Sahm et al., 2013). The industrial riboflavin synthesis using *A. gossypii* has been improved so that nowadays industrial strains and optimized process control lead to riboflavin titers of more than 20 g L^{-1} (Sahm et al., 2013). In 2002, shortly after the establishment of the *Ashbya*-process, the Swiss corporation Roche launched a riboflavin process using an engineered *Bacillus subtilis* strain, which was taken over by the Dutch company DSM a year later (DSM, 2015; Hohmann et al., 2011). The factory is located in Southern Germany in Grenzach. A new generation of riboflavin-overproducing strains do not contain multiple copies of the *rib* operon anymore, but use strong phage-derived promoters and an altered 5'-untranslated region of the *rib* operon (Sahm et al., 2013). The DSM *Bacillus*-process is also conducted in fed-batch mode, which is carbon limited. As carbon sources molasses, but also starch hydrolysates or thick juice are utilized (Kirchner et al., 2014). Corn steep liquor as well as yeast extract can be used for nitrogen supply. More recently, DSM has patented a process, in which biomass from a previous fermentation is recycled, degraded, and replaces expensive yeast extract in a new fermentation cycle (DSM, 2006). The Chinese company Hubei Guangji Pharmaceutical employs a *B. subtilis* strain that shows resistance to a proline analog to synthesize riboflavin and produces more than 26 g L^{-1} riboflavin (Lee et al., 2004). The two *B. subtilis* processes are conducted at 39 to 40 °C and last up to 70 hours (Hohmann et al., 2011; Kirchner et al., 2014; Lee et al., 2004). Another organism that was used in the large-scale industrial synthesis of riboflavin was the yeast *Candida famata* by ADM (USA). However, due to genetic instability of the overproducing mutants resulting in non-overproducing strains, the company stopped the process several years ago (Abbas and Sibirny, 2011). Recent studies demonstrated a non-reverting strain, which is able to produce about 16 g L^{-1} riboflavin during a fed-batch cultivation in a lab-scale fermenter (Dmytruk et al., 2014). Either one of the nowadays employed industrial strains has its advantages and disadvantages. The industrial process using *A. gossypii* takes long due to the separation of growth and production phase, the main advantage being that the strain is a natural overproducer. *B. subtilis*, on the other hand, is fast growing with production being coupled to growth. The recombinant nature of the *B. subtilis* overproducing strain is disadvantageous, however, and has recently become precarious as a genetically modified and unknown *B. subtilis* strain was isolated from feed-grade riboflavin imported to Europe from China. Such strains are unauthorized in the European Union (Barbau-Piednoir et al., 2015; Paracchini et al., 2017).

There are several issues that need to be considered for the large-scale production of riboflavin. Sterilization of such large volumes is carried out continuously with the media passing straight into the reaction vessel. Carbon and nitrogen sources have to be sterilized separately in order to avoid Maillard reaction. Temperatures for sterilization are kept high and exposure times short (Storhas and Metz, 2006). During the actual fermentation of riboflavin, the aeration plays a crucial role. The influence of the agitation speed and thus oxygen supply on a riboflavin fed-batch production process with *B. subtilis* was studied by a two-stage agitation speed control with the lower agitation speed (600 rpm) being beneficial during the first process phase and a higher agitation speed (900 rpm) in the later phase yielding optimal cell growth and riboflavin production (Man et al., 2014). The influence of increased aeration and the resulting growth as biofilm, as opposed to the traditional planktonic growth, was investigated for *C. famata* and found to be favorable for riboflavin biosynthesis (Mitra et al., 2012). Downstream processing of vitamin B_2 is fairly easy due to its low solubility (Figure 6). After the fermentation a large fraction of the product is already crystallized. For better separation these crystals are recrystallized with seed crystals in order to obtain cubic instead of needle-shaped crystals. Washing steps under DNA decomposing conditions (60 – 70 °C, acidic pH) are followed by separation using decantation, filtration, or centrifugation. In *A. gossypii*, the temperature increase also leads to the expression of glucanases amongst other lytic enzymes that break down the fungal cell wall (Sahm et al., 2013). For food-grade riboflavin additional washing and/or crystallization steps are necessary (Bretzel et al., 1999; Grimmer et al., 1993).

For future process improvements, whether strain, media or process parameter optimization, a reliable and parallel screening system is of utter importance. A miniaturized system, which resembles production conditions of the fed-batch operated *B. subtilis* riboflavin biosynthesis, displays an interesting, straightforward development in this direction (Knorr et al., 2007).

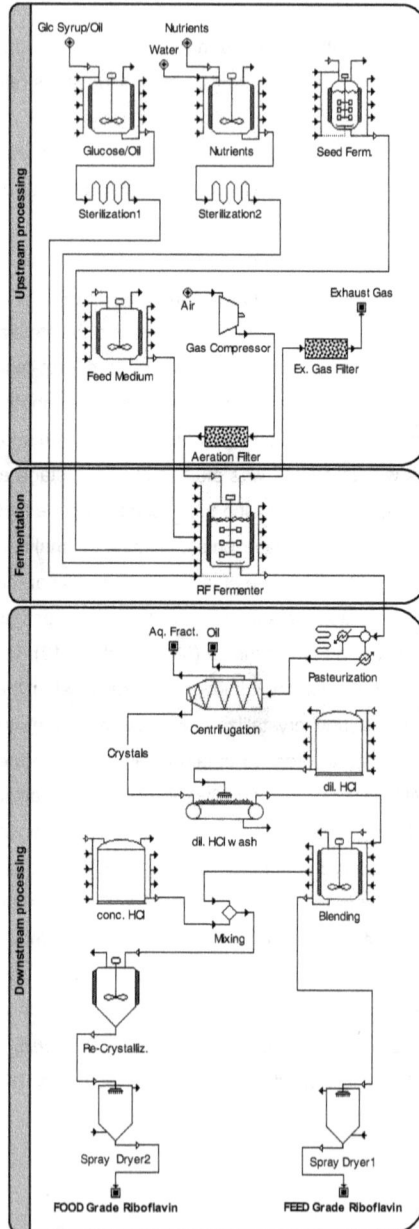

Figure 6: Flowsheet of riboflavin fermentation including upstream and downstream processing and fermentation (Schwechheimer et al., 2016).

2.5   Concept of ¹³C isotope experiments

In recent years, ¹³C tracer experiments have emerged as key technology approach to provide a better understanding of microbes (Bücker et al., 2014; Kohlstedt et al., 2014; Zamboni et al., 2004) and the knowledge derived therefrom has proven valuable to rationally engineer them (Becker et al., 2005; Buschke et al., 2011). However, already in the early 1950s, isotope tracer studies were used as means of unraveling riboflavin building blocks. The molecular origin of the carbon atoms in certain parts of riboflavin was previously clarified by tracer experiments, introducing stable ¹³C and radioactive ¹⁴C isotopes, using glucose as well as amino acids, and greatly contributed to our current knowledge (Bacher et al., 2000; Bacher et al., 1985; Bacher et al., 1998; Plaut, 1954a; Plaut, 1954b; Plaut and Broberg, 1956).

2.5.1   Conventional ¹³C metabolic flux analysis

The conventional ¹³C metabolic flux analysis (¹³C MFA) comprises an experimental and a computational part. First, a stoichiometric *in silico* model of the carbon core metabolism and other relevant metabolic reactions is derived from genome information or information about enzyme activities from the literature. Additionally, the network is expanded by the exact carbon transition between metabolites, thus providing a foundation for computational metabolite and isotopomer balancing. Experiments are then conducted using a specifically labeled ¹³C tracer substance (e.g., [1-¹³C] glucose) and defined media. The cells take up the ¹³C-labeled compound and metabolize it. Depending on the metabolic pathway used in the cell, certain metabolites will be labeled in a characteristic manner. Hereby, the different pathways will lead to specific "fingerprints". Due to their stability and abundance, proteinogenic amino acids are widely used for ¹³C labeling analysis of their respective mass isotopomer distribution (Kohlstedt et al., 2010). Due to the well-known enzymatic reaction mechanisms, carbon transitions are known and labeling of metabolic precursors can be inferred from the labeling of amino acids (Szyperski, 1995). Experimental ¹³C labeling data, as well as biomass formation, biosynthetic requirements, substrate uptake rate, and product formation rates are then integrated into the stoichiometric and isotopomer reaction model using a corresponding software. Over the years, different software tools have been developed for flux calculations from ¹³C labeling data. The open source software OpenFLUX relies on elementary mode units (EMU) and represents a particularly efficient approach regarding simulation speed (Antoniewicz et al., 2007; Quek et al., 2009). Other available software packages include FiatFlux (Zamboni et al., 2005), 13CFLUX (Wiechert et al., 2001), and a Matlab-based tool (Wittmann and Heinzle, 1999), which are all based on different approaches. Once all data are integrated using the software, free fluxes are simulated and labeling distributions are iteratively calculated, until the deviation between simulated and experimental data is minimal. This type of ¹³C metabolic flux analysis requires metabolic and isotopic steady-state, which is usually the case during exponential growth (Wittmann, 2007).

Over the past decade ^{13}C MFA has proven to be a powerful tool when it comes to gaining deeper insights into cellular metabolism and understanding underlying mechanisms (Becker et al., 2005; Buschke et al., 2011; Kohlstedt et al., 2014; Zamboni et al., 2004). Most recently, succinate production from sucrose with the rumen bacterium *Basfia succiniciproducens* was explored and optimized using four parallel tracer conditions. Findings regarding the sucrose metabolism gained from the ^{13}C MFA were directly applied by creating a mutant strain with a deleted fructose phosphotransferase system (PTS), yielding a strain with improved production performance (Lange et al., 2017). A great example for metabolic pathway engineering based on ^{13}C MFA is lysine production by *C. glutamicum*, where the combination of several selected targets led to the *de novo* generation of an industrially competitive hyperproducer (Becker et al., 2011). Rational metabolic engineering of microorganisms displays an important application for ^{13}C MFA, however, it also provides better understanding of the metabolism in general. With the help of ^{13}C MFA, the pyruvate-tricarboxylic acid (TCA) cycle node could be identified as crucial point for controlling virulence in *Yersinia pseudotuberculosis* (Bücker et al., 2014). Subsequent deletion of the gene coding for pyruvate kinase rendered strains with reduced virulence. This traditional ^{13}C MFA approach, however, is only applicable for well-defined microbial processes that entail chemically defined media with one carbon source as well as isotopic and metabolic steady-state and preferably parallel growth and production.

2.5.2 Metabolic flux studies under complex conditions

Metabolic flux studies were, however, also conducted in rather complex scenarios that displayed characteristics not generally applied for ^{13}C MFA: undefined and complex media, multiple carbon sources, microbial consortia, and instationary cultivation conditions. Perfused rat liver models were used to investigate the TCA cycle and gluconeogenic fluxes by tracing the hepatic glutamate labeling after perfusion with ^{13}C-labeled lactate (Large et al., 1997). The study proved the validity of a model proposed by Magnusson et al. (1991), who investigated liver metabolism in humans using a set of equations considering carbon transitions in the TCA cycle and gluconeogenesis. In the field of biotechnology, recent studies have taken the ^{13}C MFA approach further in order to investigate the rather complex cocoa pulp fermentation. Parallel labeling experiments using ^{13}C-labeled sugars in an otherwise rather undefined cocoa pulp simulation medium were conducted using lactic acid bacteria strains (Adler et al., 2013). Metabolic fluxes of pure cultures as well as microbial consortia were calculated and revealed the different levels of contribution to the cocoa fermentation of single strains. *Lactobacillus fermentum* NCC 575 could be identified as a dominant contributor to the overall carbon flux in the chosen set-up (Adler et al., 2013). In a subsequent study, the authors further investigated the cocoa fermentation with regard to consumption of a substrate mixture and carbon fluxes in acetic acid bacteria (Adler et al., 2014). With the help of parallel ^{13}C tracer experiments, lactate could be identified as main substrate for biomass and the product acetoin, whereas ethanol contributed mainly to the product acetate, which is an important precursor for characteristic cocoa flavor compounds. This functionally separated metabolism in acetic acid bacteria could be attributed to the lack of two enzymes at the pyruvate node, namely phosphoenolpyruvate (PEP) carboxykinase and malic enzyme (Adler et al., 2014). In combination, the experiments successfully resolved the complex process of cocoa fermentation using complementary ^{13}C tracer studies.

2.5.3 ^{13}C Labeling analysis

The common feature of the above-mentioned ^{13}C labeling studies (Chapters 2.5.1 and 2.5.2) is the use of GC/MS to quantify mass isotopomer distributions of proteinogenic amino acids or other metabolites. This very robust method provides highly accurate measurements with excellent separation of complex reaction mixtures (Pan and Raftery, 2007; Wittmann, 2007). The high sensitivity of this technique enables the measurement of even less abundant metabolites (Dersch et al., 2016). The ^{13}C labeling analysis with GC/MS yields information about the mass isotopomer distribution of a certain analyte, thus giving information about the number of ^{13}C atoms in a certain molecule, but not their respective positions in the molecule (Figure 7). However, the information content can be increased by measuring different ion cluster fragments of the same molecule (Wiechert, 2001). GC/MS is only applicable for volatile and thermally stable compounds. In case an analyte does not meet those demands, derivatization is required, such as silylation of amino acids using *N*-methyl-*N*-*tert*-butyldimethylsilyl-trifluoroacetamide (MBDSTFA) (Pan and Raftery, 2007; Wiechert, 2001; Wittmann, 2007). LC/MS shares many advantages of GC/MS, however, derivatization of analytes is not necessary (Pan and Raftery, 2007; Wiechert, 2001), thereby eliminating an additional sample preparation step (Figure 7). The soft ionization method (electrospray ionization) renders fully intact molecules (Alder et al., 2006), however, the use of a tandem mass spectrometry supplies more information on ^{13}C labeling due to the fragmentation of the analytes in a collision cell. High salt concentrations of fermentation media might interfere with the electrospray ionization source (Mashego et al., 2007), thus desalination of samples prior to measurement could be necessary. While the MS-based techniques typically display high sensitivity (Figure 7), NMR exhibits lower sensitivity, therefore, requiring higher metabolite concentrations (Barding et al., 2013). However, NMR provides data on positional enrichment of the analyte as well as structural information (Figure 7), which offers the greatest information content compared to the other two techniques (Barding et al., 2013; Wiechert, 2001). In addition, the method is advantageous for compounds that are difficult to ionize and no elaborate sample preparation is necessary (Markley et al., 2017). The information content of NMR can be increased even further when multinuclear and multidimensional NMR are applied (Dersch et al., 2016), however a skilled specialist is required to interpret the resulting complex spectra (Wiechert, 2001). Since even a single analyte containing several carbon atoms can give rise to a highly sophisticated spectrum, complex samples can lead to resonance overlap of different compounds (Barding et al., 2013). Each of the described techniques is characterized by a unique set of advantages and disadvantages. However, only the custom-made combination of all methods and integration of the resulting data will provide the highest information content, especially for highly complex microbial processes.

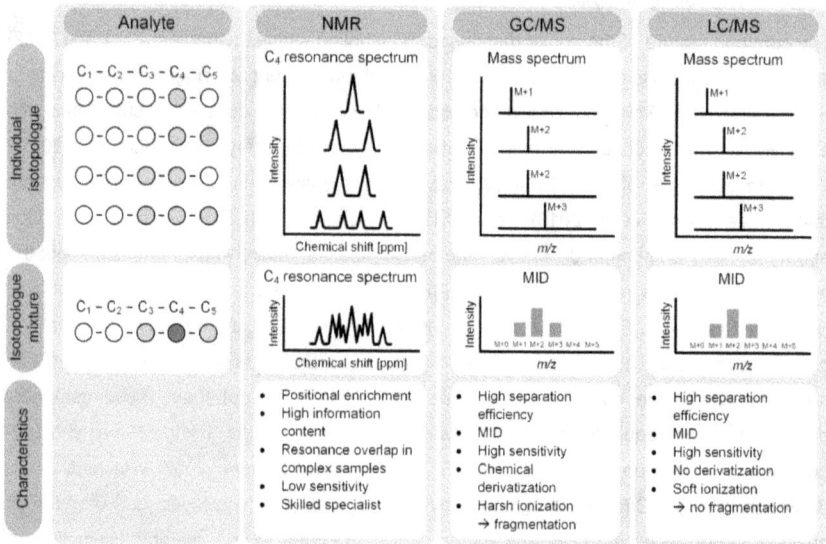

Figure 7: Information content and characterization of different analytical methods for ^{13}C labeling analysis for metabolic flux analysis in microorganisms, i.e. nuclear magnetic resonance (NMR), gas chromatography/mass spectrometry (GC/MS), and liquid chromatography/mass spectrometry (LC/MS). As example, different isotopologues of a molecule containing five carbon atoms and their respective mixture are depicted. The resulting spectra reveal that NMR provides information about the positional enrichment of each carbon atom, whereas GC/MS and LC/MS usually yield information regarding the mass isotopomer distribution (MID), but is unable to differentiate between positional isomers. However, the combined information of different ion cluster fragments can substantially enhance the information content for GC/MS or LC/MS samples.

2.5.4 Flux analysis combining different analytical techniques

The successful combination of different analytical methods for ^{13}C labeling analysis has proven to be a powerful tool in several studies. The presence of pyruvate carboxylating activity in *C. glutamicum* could be identified and demonstrated in a ^{13}C tracer study (Park et al., 1997). The ^{13}C labeling of extracellular lysine was monitored using GC/MS as well NMR, rendering high resolution data. In a study using glycerol overproducing *S. cerevisiae*, the combined results of GC/MS, LC/MS, and NMR led to an accurately resolved flux distribution with especially precise fluxes around important metabolic nodes (Kleijn et al., 2007). The data provided important evidence that carbon was directed away from glycerol formation to the PP pathway as well as assimilatory pathways towards storage compounds. The complementary nature of GC/MS and LC/MS was applied for instationary ^{13}C MFA using *Pichia pastoris* (Jordà et al., 2013). Methanol and glucose were co-metabolized and the use of LC/MS and GC/MS supported the simulation of a large metabolic network including trehalose recycling as well as reversibilities of the glycolytic/gluconeogenic pathway, the TCA cycle, and PP pathway (Jordà et al., 2013). In addition the ratio of methanol oxidation versus methanol utilization could be determined in a precise manner. Schatschneider et al. (2014) conducted ^{13}C MFA with *Xanthomonas campestris* pv. campestris on minimal medium with ^{13}C-labeled glucose. GC/MS was applied for detection of ^{13}C labeling in proteinogenic amino acids and subsequent flux simulations, which confirmed the role of the Entner-Doudoroff pathway as main pathway for glucose utilization. However, only ^{13}C NMR-based isotopologue profiling of amino acids was able to resolve glucose-utilizing pathways in greater depth (Schatschneider et al., 2014).

2.5.5 Concept of ^{13}C isotope tracer studies with *A. gossypii*

The industrial riboflavin production with *A. gossypii* comprises a highly undefined culture medium with large amounts of yeast extract as well as a main carbon source, rapeseed oil, which is commercially not available as ^{13}C-labeled substrate. In addition, growth and riboflavin production phase are separated (Sahm et al., 2013). Thus, this set-up does not meet the criteria needed for conventional ^{13}C MFA as described above (Chapter 2.5.1). However, studies with challenging microbial systems such as the cocoa fermentation as well as the great resolution obtained by combination of complementary analytical techniques (GC/MS, LC/MS, NMR) as described above (Chapters 2.5.2 - 2.5.4) provide an excellent starting point for the ^{13}C tracer-based resolution of industrial riboflavin production with *A. gossypii*.

In order to resolve carbon fluxes of the riboflavin overproducer *A. gossypii* using a complex medium with rapeseed oil, a sophisticated approach was developed and then applied (Figure 8). In the first step, a suitable ^{13}C tracer was selected (e.g. [$^{13}C_2$] glycine). The labeled tracer substituted the naturally labeled tracer in the medium. Cultivation also required non-labeled replicates in order to obtain information on biomass formation, substrate consumption, product formation, and yields.

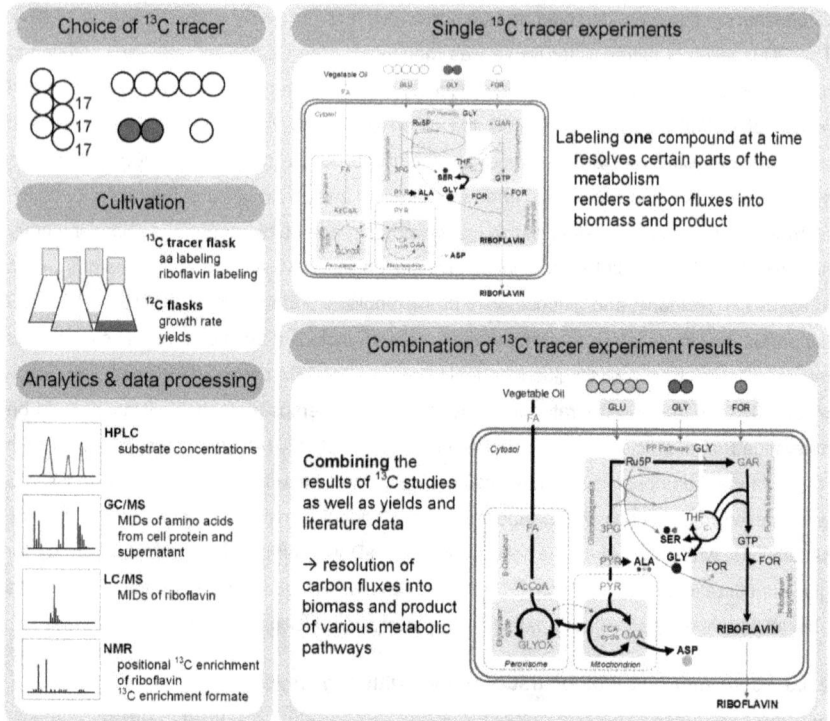

Figure 8: Sophisticated set-up of ^{13}C tracer studies for resolving carbon fluxes of the riboflavin overproducer *A. gossypii*. At first, a suitable ^{13}C tracer is chosen, which substitutes the naturally labeled compound in the medium. Cultivation is carried out using the ^{13}C labeled culture, which will be analyzed for amino acid and riboflavin labeling, as well as unlabeled replicates, which are used in order to derive yields and rates. The analytical spectrum comprises high performance liquid chromatography (HPLC) as means of quantifying substrate concentrations, gas chromatography/mass spectrometry (GC/MS) in order to gain information about mass isotopomer distributions (MID) for amino acids (proteinogenic and from the supernatant), liquid chromatography/mass spectrometry (LC/MS) for MIDs of riboflavin, ^{13}C NMR in order to gain ^{13}C positional enrichment information of riboflavin, and ^{1}H NMR for ^{13}C enrichment of formate and determination of formate concentration. Once all data are processed and implemented, qualitative carbon fluxes for a single experiment can be obtained. This includes fluxes for only certain parts of the metabolism. The whole workflow is then repeated using a different ^{13}C tracer, which will render complementary results. In the end, combining the results of all ^{13}C tracer studies will provide a thorough picture of qualitative and quantitative fluxes into biomass and product formation.

The analytical spectrum comprised high performance liquid chromatography (HPLC) analyses for the determination of substrates. For identification of mass isotopomer distributions of amino acids, the conventional GC/MS measurement could be used. Riboflavin, which cannot be analyzed via GC/MS, was measured using LC/MS. Due to the large amount of carbon atoms, single ^{13}C incorporation effects were difficult to resolve with high precision. Thus, ^{13}C NMR was also applied, which is able to provide data on ^{13}C enrichment of single carbon atoms. Finally, ^1H NMR was the method of choice for formate quantification as well as ^{13}C enrichment measurement. This complex experimental workflow was then repeated for a different ^{13}C tracer. In the end, the combination of the results from the individual ^{13}C tracer experiments, will allow the resolution of metabolic fluxes for this complex system.

3 MATERIAL AND METHODS

3.1 Strains

All strains were obtained from the BASF (Ludwigshafen, Germany): *Ashbya gossypii* WT, the wild type strain (ATCC 10895) and the riboflavin overproducing *A. gossypii* B2 (derived from the wild type strain). Strains were kept as glycerol stocks at -80 °C.

3.2 Chemicals and media

3.2.1 Chemicals

All chemicals were purchased from Sigma-Aldrich (Taufkirchen, Germany), Roth (Karlsruhe, Germany), Becton Dickinson (Franklin Lakes, NJ, USA), and Merck (Darmstadt, Germany) at analytical grade quality, if not stated otherwise. Rapeseed oil, soy flour, yeast extract, and corn swell flour were obtained from BASF. Stable ^{13}C tracers used in this work are listed in Table 3. Water was purified by a Milli-Q Integral water purification system (Merck).

3.2.2 Medium composition

Cultivation of *A. gossypii* on agar plates was performed using sporulation (SP) agar plates, which contained per liter: 3 g soy flour, 3 g yeast extract, 3 g malt extract, 20 g corn swell flour, 30 g agar, and 10 g glucose. A stock solution of glucose (400 g L^{-1}) was autoclaved separately before addition to SP medium. The pH was adjusted to 6.8.

The pre-culture medium consisted of 57.9 g L^{-1} yeast extract, 0.5 g L^{-1} $MgSO_4 \cdot 7H_2O$, 40 g L^{-1} rapeseed oil (pH 7). For glycerol stocks 9.5 % (v v^{-1}) glycerol was added to the pre-culture medium. Medium for the main culture contained per liter: 28.6 g yeast extract, 9.5 g glycine, 7.4 g sodium glutamate-monohydrate, 1.9 g KH_2PO_4, 0.5 g $MgSO_4 \cdot 7H_2O$, 1.1 g L-methionine, 0.2 g *m*-inositol, 2.0 g sodium formate, 9.0 g urea, and 156.5 g rapeseed oil (pH 7). Urea and rapeseed oil were sterilized separately by filtration and autoclaving, respectively. For cultivations using glucose as main carbon source, rapeseed oil was replaced with a final concentration of 20 g L^{-1} glucose (111 mM or 666 C-mM). In addition, the medium was buffered with *N*-(2-acetamido)-2-aminoethanesulfonic acid (ACES) (200 mM). Cultivations on glycerol and sodium acetate were carried out with the same combined C-molarity (666 C-mM) in a C-molar ratio of 1:27 for glycerol and acetate, assuming the main component of rapeseed oil being C18 fatty acids, which form esters with glycerol. Medium using glycerol and acetate as carbon source was buffered with 250 mM 3-(*N*-morpholino)propanesulfonic acid (MOPS).

3.3 Cultivation

3.3.1 Shake flask cultivation

Mycelia from cryo cultures were incubated for five to seven days on SP agar plates at 30 °C. With mycelia from those agar plates a pre-culture was inoculated using non baffled shake flasks with 10 % filling volume (40 h). All cultivations were conducted on a rotary shaker at 200 rpm, 30 °C, and 80 % humidity (Multitron Pro, Infors, Bottmingen, Switzerland). The main culture was inoculated from the pre-culture with an initial inoculum size of 3.5 % (v v^{-1}) and was carried out in 250 mL baffled shake flasks with a filling volume of 30 mL (144 h) (Figure 9A). Cultivations were performed in parallel shake flasks with three flasks being sacrificed per sample (Figure 9). This was necessary in order to obtain reliable and reproducible data. Firstly, this enabled large sampling volumes in the early growth phase, required for accurate cell dry weight (CDW) determination. Secondly, this avoided potential sampling artefacts, otherwise resulting from changes in oil distribution and phase separation between aqueous and oil phase throughout the course of the experiments. Eventual evaporation was monitored throughout the cultivation by weighing the shake flask before and after each sample. All concentrations were corrected for the rate of evaporation.

Table 3: Isotopic tracers used in this work. Stable ^{13}C isotopes were used for qualitative ^{13}C tracer experiments.

Tracer	Supplyer	Labeling Purity [%]	Reference
[^{13}C$_6$] Glucose	Omicron[a]	99	Chapter 4.2
[^{13}C$_2$] Sodium acetate	Sigma-Aldrich[b]	99	Chapter 4.3
[^{13}C$_3$] Glycerol	Sigma-Aldrich[b]	99	Chapter 4.3
[^{13}C$_2$] Glycine	Sigma-Aldrich[b]	99	Chapters 4.4, 4.5
[^{13}C$_3$] Serine	Sigma-Aldrich[b]	99	Chapters 4.4, 4.5
[3-^{13}C] Serine	Sigma-Aldrich[b]	99	Chapter 4.5
[^{13}C] Sodium formate	Sigma-Aldrich[b]	99	Chapters 4.4, 4.5
[U^{13}C] Yeast extract	Ohly[c]	99	Chapters 4.4, 4.5
[^{13}C$_5$] Glutamic acid	Sigma-Aldrich[b]	98	Chapters 4.4, 4.5

[a] South Bend, IN, USA
[b] Taufkirchen, Germany
[c] Hamburg Germany

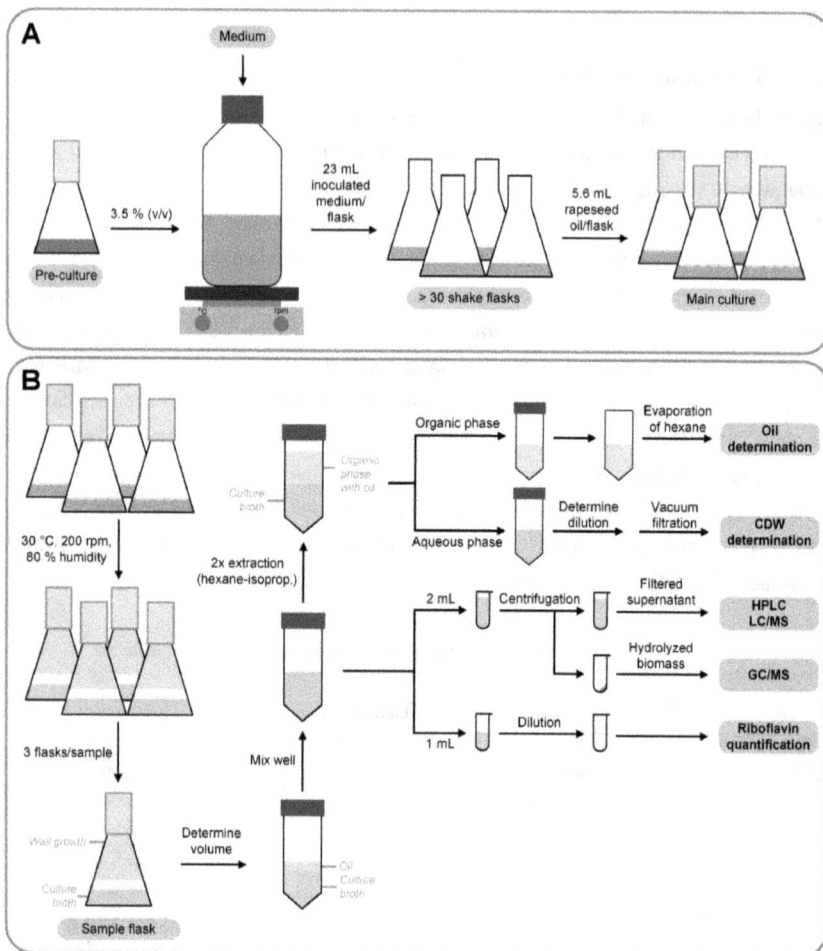

Figure 9: Inoculation (A) and sampling scheme (B) of standard cultivations with *A. gossypii*, using rapeseed oil as main carbon source for naturally labeled substrates.

3.3.2 Cultivations with ^{13}C tracer compounds

In isotope experiments, naturally labeled ingredients were replaced by ^{13}C labeled tracers in equimolar amount or were added to the medium (Table 3). Natural glycine was replaced by $[^{13}C_2]$ glycine. Natural sodium glutamate was replaced by $[^{13}C_5]$ glutamic acid. In additional experiments, $[^{13}C_3]$ serine as well as $[3-^{13}C]$ serine were added to a tracer concentration of 2.4 g L^{-1} at different time points of the cultivation (0 h or 48 h). Natural formate was either replaced in the initial medium by [^{13}C] sodium formate or was added after 48 h to a tracer concentration of 2.0 g L^{-1}. In the latter case, the initial medium contained the normal amount of natural formate given above. The company Ohly (Hamburg, Germany) provided [U^{13}C] yeast extract, which replaced the naturally labeled yeast extract. All other ^{13}C enriched compounds were obtained from Sigma-Aldrich (Taufkirchen, Germany) or Omicron (South Bend, Indiana, USA). All the ^{13}C tracer studies conducted in this work are listed in the appendix (Table 15).

3.3.3 Supplementation studies

Formate supplementation studies were done using main culture medium without formate in the initial medium and standard cultivation conditions. Sodium formate was then added after 0 h, 12 h, 24 h, and 36 h in parallel experiments (final concentration 2.0 g L^{-1}). Three flasks per condition were harvested after 144 h and the riboflavin titer was determined. Supplementation studies using serine were conducted in a similar manner. Medium without formate was supplemented with L-serine (final concentration 2.4 g L^{-1}) at different time points (0 h, 12 h, 24 h, 36 h) in parallel cultivations. The riboflavin titer was measured after 144 h. In both supplementation cases, the original medium (with formate in the initial medium) served as control.

3.3.4 Sampling of biomass and supernatant

Due to the unconventional nature of the cultivation set-up, the sampling scheme had to be designed carefully. Figure 9B shows the workflow for the naturally labeled control flasks for the determination of yields and rates. Sampling of the ^{13}C tracer condition had to be adjusted: 1 mL to 2 mL culture broth were sampled per time point from the same flask throughout the cultivation. The culture broth was then centrifuged. The supernatant and pellet were used for GC/MS, LC/MS, and HPLC analysis (Figure 9B).

3.3.5 Sampling for intracellular metabolites

Samples for intracellular amino acid and formate extraction were obtained by fast filtration as described elsewhere (Bolten and Wittmann, 2008). Briefly, 0.7 mL cell culture was vacuum filtered (Whatman filter paper, Grade 3, GE Healthcare, Little Chalfont, UK) and washed with medium without amino acids or formate. For metabolite extraction, the harvested cells were then incubated in 5 mL α-aminobutyric acid (200 µM) at 100 °C. After 15 min, extracts were cooled, transferred to a reaction tube, centrifuged (centrifuge 5415R, Eppendorf, Hamburg, Germany), and filtered (0.2 µm, PES, Sarstedt, Nümbrecht, Germany).

3.4 Analytical methods

3.4.1 Cell concentration

Cell dry weight (CDW) was determined gravimetrically by filtration of culture broth (Whatman filter paper Grade 3, GE Healthcare). Residual rapeseed oil was removed as described below (Figure 9B). The remaining aqueous phase was filtered. Cells and filter were dried and weighed using a moisture analyzer (HB43-S, Mettler Toledo, Columbus, OH, USA) (Stahmann et al., 1994). Dilution effects of the preceding extraction step using hexane-isopropanol were taken into account.

3.4.2 Quantification of substrates and products

3.4.2.1 Gravimetric oil determination

Preceding CDW determination, residual oil in the flask was also measured gravimetrically by extracting the culture broth twice with one volume hexane-isopropanol (3:2 v v^{-1}). The upper layer containing rapeseed oil was collected and the remaining solvent was evaporated until a constant weight was reached (Figure 9B) (Stahmann et al., 1994).

3.4.2.2 Riboflavin

Quantification of riboflavin was carried out spectrophotometrically at a wavelength of 444 nm (UV-1600PC spectrophotometer, VWR, Hannover, Germany). Samples were diluted using 1 N NaOH and neutralized by adding 0.1 M potassium phosphate buffer (pH 6) (Jeong et al., 2015).

3.4.2.3 Amino acids

Amino acids in the culture supernatant as well as in metabolite extract were quantified using HPLC (Agilent 1200 series, Agilent Technologies, Waldbronn, Germany) with α-aminobutyric acid as internal standard and pre-column derivatization (Krömer et al., 2005). The separation was conducted using a reversed-phase column (Gemini 5 µm C18 110 Å, 150 x 4.6 mm, Phenomenex, Aschaffenburg, Germany) and gradient elution (0 min: 100 % eluent A; 44.5 min: 55.5 % eluent A; 45.0 min: 39 % eluent A; 47 min: 39 % eluent A; 48 min: 18 % eluent A; 48.5 min: 0 % eluent A; 50.5 min: 0 % eluent A; 51 min: 100 % eluent A; 53 min: 100 % eluent A). The amino acid derivatives were detected via fluorescence (excitation at 340 nm, emission at 540 nm). All samples were filtered prior to measurement (0.2 µm, PES, Sarstedt, Nümbrecht, Germany).

3.4.2.4 Formate

Formate quantification in supernatant and in metabolite extract was conducted, using ^{1}H NMR. Culture supernatants and extracts were filtered (0.2 µm) and the internal standard 2-

(trimethylsilyl)propionic-2,2,3,3-d4 acid (TSP-d4) was added. The measurement was carried out on a Bruker Ascend Avance III 800 MHz spectrometer (Bruker, Rheinstetten, Germany) using the Bruker TopSpin software for data analysis (Ryffel et al., 2016) and at 293 K with a total of 16 scans at a flip angle of 30 ° (10 s relaxation delay, 7.53 μs and -12.55 dB hard pulse (90 °)). The spectral window width of the measurement was 20 ppm. Following changes were made for formate quantification from metabolite extracts: 128 scans at 298 K and 7.57 μs duration of hard pulse.

3.4.3 Mass isotopomer distributions

3.4.3.1 Mass isotopomers of amino acids

Mass isotopomer distributions (MIDs) of amino acids from hydrolyzed cell protein and culture supernatant were quantified by GC/MS using helium as carrier gas (Gas chromatograph 7890B, Agilent; mass selective detector 5875A, Agilent; column: HP-5-MS, 30 m x 0.25 mm x 0.25 μm, Agilent) (Kiefer et al., 2004). Cells were harvested at the end of the exponential growth phase (32 to 36 h for cultivations on vegetable oil, 12 to 24 h for cultivations on other main carbon sources). The culture broth was centrifuged. The collected pellet was washed several times with deionized water, followed by hydrolysis with 6 M HCl (100 °C, 24 h). Hydrolysates were clarified from cell debris (0.2 μm, Merck Millipore, Darmstadt, Germany), dried under nitrogen, and derivatized using 0.1 % pyridine in dimethylformamide and N-methyl-N-tert-butyldimethylsilyl-trifluoroacetamide (MBDSTFA) as previously described (Wittmann et al., 2002) (Figure 10A). Culture supernatant was filtered (0.2 μm) and an amount, containing about 30 nmol of the amino acids of interest, was dried under nitrogen and derivatized as mentioned above. Separation of the silylated amino acids was carried out using the following temperature profile: 2 min/120 °C, temperature increase to 200 °C at a rate of 8 °C/min, temperature increase to 323 °C at a rate of 10 °C/min. The temperature of the inlet, the interface, and the quadrupole was set to 280 °C. All samples were first measured in scan mode to check for isobaric interference with the analytes of interest and verify unbiased ion clusters for ^{13}C quantification based on retention time and unique fragmentation pattern. Subsequently, labeling patterns of all samples were determined in technical duplicates using selective ion monitoring (SIM) of selected ion clusters. A GC/MS spectrum of tert-butyl-dimethylsilyl-derivatized proteinogenic amino acids of A. gossypii grown on vegetable oil in SIM mode is shown in Figure 10C.

The ion cluster [M-57] was measured for all amino acids, but leucine and isoleucine, as it represents the full carbon backbone of the analytes (Wittmann, 2007). The ion cluster [M-85] had to be measured for proline, as its [M-57] fragment overlapped with succinate. The [M-85] cluster was also analyzed for isoleucine and leucine as their [M-57] ion cluster represents both the full carbon backbone as well as the [M-residue] fragment, which contains solely carbon atoms 1 and 2 (Figure 10B). The mass fragments used for further analysis are listed in Table 4. The MIDs obtained from the measurements had to be corrected for abundance of natural isotopes and dilution of labeling through naturally labeled pre-culture as described in Chapter 3.5.1.

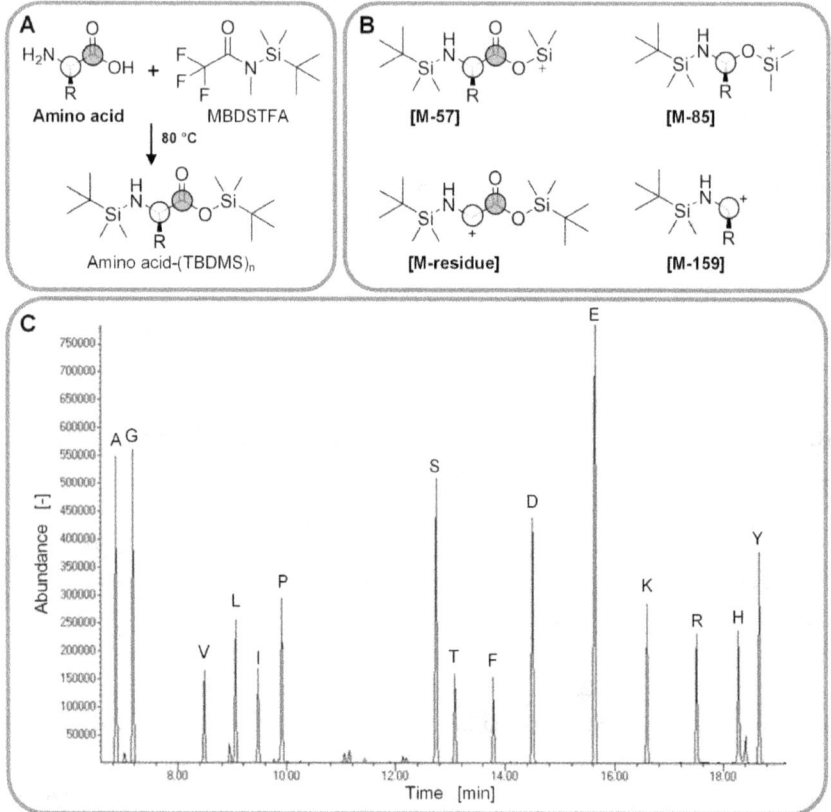

Figure 10: Derivatization of amino acids (A), subsequent fragmentation of *tert*-butyl-dimethylsilyl-derivatized amino acids during GC/MS analysis (B), and GC/MS spectrum in SIM mode of TBDMS-derivatized proteinogenic amino acids of *A. gossypii* B2 grown on complex medium with rapeseed oil as main carbon source (C). Amino acids were derivatized using MBDSTFA at 80 °C, rendering the corresponding TBDMS-derivative. Grey circles denote the carbon atom at position 1, white circles correspond to position 2. The subscript n designates the number of functional groups that were silylated in the respective amino acid (A). During electron ionization derivatized amino acids are fragmented yielding characteristic ion clusters (B). R, residues of the amino acid. Letters in (C) depict the one-letter code for amino acids.

Table 4: Fragment ion clusters of TBDMS derivatives of amino acids and the TMS derivative of glucose measured by GC/MS in SIM mode. The mass given for each analyte, M+0, represents the mass-to-charge ratio (m/z) of the isotopomer that contains only non-labeled atoms. TBDMS, *tert*-butyl-dimethylsilyl; (TMS)$_5$-MOA-glucose, 5-trimethylsilyl-O-methyloxime glucose; residue, residue of amino acid.

Analyte	M+0 [m/z]	Fragment	Carbon atoms
Alanine-(TBDMS)$_2$	260	[M-57]	1 – 3
	232	[M-85]	2 – 3
Glycine-(TBDMS)$_2$	246	[M-57]	1 – 2
	218	[M-85]	2
Valine-(TBDMS)$_2$	288	[M-57]	1 – 5
	260	[M-85]	2 – 5
Leucine-(TBDMS)$_2$	274	[M-85]	2 – 6
Isoleucine-(TBDMS)$_2$	274	[M-85]	2 – 6
Proline-(TBDMS)$_2$	286	[M-57]	1 – 5
	258	[M-85]	2 – 5
Serine-(TBDMS)$_3$	390	[M-57]	1 – 3
	362	[M-85]	2 – 3
Threonine-(TBDMS)$_3$	404	[M-57]	1 – 4
	376	[M-85]	2 – 4
Phenylalanine-(TBDMS)$_2$	336	[M-57]	1 – 9
	234	[M-159]	2 – 9
Aspartate-(TBDMS)$_3$	418	[M-57]	1 – 4
	390	[M-85]	2 – 4
Glutamate-(TBDMS)$_3$	432	[M-57]	1 – 5
	330	[M-159]	2 – 5
Lysine-(TBDMS)$_3$	431	[M-57]	1 – 6
	329	[M-85]	2 – 6
Arginine-(TBDMS)$_3$	442	[M-57]	1 – 6
Histidine-(TBDMS)$_2$	440	[M-57]	1 – 6
Tyrosine-(TBDMS)$_3$	466	[M-57]	1 – 9
	302	[M-residue]	1 – 2
(TMS)$_5$-MOA-Glucose	544	[M-15]	1 – 6

3.4.3.2 Mass isotopomers of glycogen-derived glucose by GC/MS

The mass isotopomer distribution for glucose from cellular glycogen was measured with the same GC/MS system as described above (see Chapter 3.4.3.1). About 5 mg cells were harvested at the end of the exponential growth phase by centrifugation. Recovery of glycogen from biomass was carried out as follows: harvested cells were washed several times with deionized water and resuspended in 500 µL enzyme solution, containing 70 U mL^{-1} amyloglucosidase (Sigma-Aldrich) and 1.2 kU mL^{-1} lysozyme (0.02 M Tris HCl, pH 7.0) (Sigma-Aldrich). Subsequent hydrolysis occurred at 37 °C for 3 h at 400 rpm. After cell debris removal via filtration (0.2 µm, Merck Millipore), the glucose concentration was determined enzymatically, using a D-glucose assay kit according to instructions by the manufacturer (R-Biopharm, Darmstadt, Germany). At least 40 µg glucose were then dried by lyophilization, followed by a two-step derivatization using first 50 µL 2 % methoxylamine in pyridine (80 °C, 25 min) and then adding 50 µL N,O-bis-trimethylsilyl-trifluoroacetamide (BSTFA) (80 °C, 30 min) (Figure 11A). Measurement of the obtained trimethylsilyl O-methyloxime derivative of glucose was carried out using the following temperature profile: 3 min/150 °C, temperature increase to 230 °C at a rate of 8 °C/min, temperature increase to 325 °C at a rate of 25 °C/min. The temperature of the inlet, the interface, and the quadrupole was set to 280 °C. All samples were first measured in scan mode to check for isobaric interference. The mass isotopomer distribution was then determined in technical duplicates using SIM of the ion cluster [M-15], which represents all six carbon atoms of glucose (Figure 11B, Table 4) (Laine and Sweeley, 1971). The resulting MIDs were corrected as described below (Chapter 3.5.1).

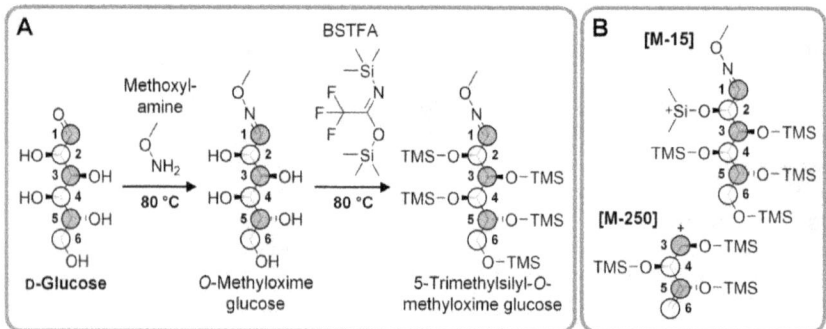

Figure 11: Two- step derivatization (A) of glycogen-derived cellular glucose using methoxylamine and BSTFA. GC/MS fragmentation (B) renders a [M-15] fragment containing all six carbon atoms of glucose after loss of a methyl group and the [M-250] fragment containing carbon atoms 3 to 6.

3.4.3.3 Mass isotopomers of riboflavin

The ^{13}C labeling of riboflavin in the culture supernatant was analyzed by LC/MS (Hoffmann et al., 2014). Samples were filtered twice (0.2 µm) before treatment with pipet-tip columns of C18 resin (Pierce C18 tips, 100 µL bed size, ThermoFisher Scientific, Dreieich, Germany). All samples were analyzed using a Dionex Ultimate 3000 RSLC (ThermoFisher Scientific) and a maXis 4G hr-QqToF mass spectrometer (Bruker, Rheinstetten, Germany) using the Apollo ESI source. Settings for LC-separation were as described previously (Hoffmann et al., 2014) with a modified gradient: 0 min: 5 % eluent B; 0.5 min: 5 % eluent B; 9.5 min: 95% eluent B; 10.5 min: 95 % eluent B; 11.0 min: 5 % eluent B; 12.5 min: 5 % eluent B. With a few exceptions, mass spectrometric parameters were kept unchanged to the previously described method (Hoffmann et al., 2014): quadrupole ion energy was set to 5 eV with a low mass of 120 m/z. The collision cell was also set to 5 eV with a collision RF of 800 Vpp. Obtained MIDs for the 377 m/z parent ion of riboflavin were corrected as described below (Chapter 3.5.2).

3.4.4 Positional isotopomers

3.4.4.1 Positional isotopomers of riboflavin

Positional labeling data for riboflavin were obtained by ^{1}H and ^{13}C NMR. For NMR analyses of riboflavin, rapeseed oil of the culture was extracted, cells and riboflavin crystals were filtered using Whatman filter paper (GE Healthcare). The filter cake was resuspended in 80 mM NaOH. Upon cell removal by centrifugation and filtration (0.2 µm), riboflavin was precipitated by addition of 80 mM HCl. The collected pellet was lyophilized and then dissolved in dimethyl sulfoxide for 48 h to a final concentration of 2 g L^{-1}. NMR spectra were recorded on a Bruker Ascend Avance III 800 MHz spectrometer (Bruker, Rheinstetten, Germany) using the Bruker TopSpin software for data analysis (Ryffel et al., 2016). The ^{1}H NMR measurements were carried out at 800 MHz at a nominal temperature of 300 K with a total of 4 scans at a flip angle of 30 ° (5 s relaxation delay, 7.31 µs and -12.55 dB hard pulse (90 °)). For every sample, 64k free induction decay (FID) signals were collected with a spectral width of 20 ppm. All ^{13}C NMR measurements were conducted at 201.2 MHz with a total of 2304 scans and a spectral width of 236.6 ppm at 6 s relaxation delay, 12.3 µs and -23.01 dB hard pulse (90 °). To account for the exact contribution of the used tracer to the synthesis of a specific riboflavin carbon atom, the relative ^{13}C enrichment of this atom was corrected (Chapter 3.5.2).

3.4.4.2 Quantification of ^{13}C enrichment of formate

Culture supernatants were filtered (0.2 µm) and the internal standard 2-(trimethylsilyl)propionic-2,2,3,3-d4 acid (TSP-d4) was added. The ^{13}C enrichment of formate was measured indirectly via ^{1}H NMR, which was carried out at 293 K with a total of 16 scans at a flip angle of 30 ° at 10 s relaxation delay, 7.53 µs and -12.55 dB hard pulse (90 °). The spectral window width of the measurement was 20 ppm.

3.5 Correction of labeling data

3.5.1 Correction of ^{13}C labeling data from hydrolyzed cell protein and glycogen

All MIDs were corrected for natural isotopes (van Winden et al., 2002) as well as for the fraction of naturally labeled amino acids originating from the inoculum. The output of the correction was a mass isotopomer distribution of the carbon skeleton of the analytes formed during the corresponding isotope experiment (MID_{corr}). The total ^{13}C enrichment of the carbon skeleton of a compound with n carbon atoms, termed summed fractional labeling (SFL) (Christensen et al., 2000), was calculated according to Equation (1). The data were given in percent, whereby 100 % represents a fully ^{13}C-labeled carbon backbone and i represents the number of ^{13}C atoms of this isotopomer.

$$SFL = \sum_{i=1}^{n+1} \frac{i \cdot MID_{i,corr}}{n} \cdot 100 \qquad \text{(Eq. 1)}$$

To calculate the net incorporation of ^{13}C label from the tracers glucose, acetate, and glycerol into the respective proteinogenic amino acids, the SFL was then corrected for the natural ^{13}C background of the carbon backbone (1.07 %) (Adler et al., 2014) and the purity P of the respective tracer (Table 3) according to Equation (2).

$$SFL_{corr} = \frac{SFL - SFL_{nat}}{P} \cdot 100 \qquad \text{(Eq. 2)}$$

For all other labeling studies, the SFL was corrected for the natural ^{13}C background of the carbon backbone as well as fraction of labeling originating from unlabeled pre-culture of the respective tracer at the beginning of the cultivation (appendix Figure 39). Equation (3) then yielded the corrected SFL$_{corr}$ (all SFLs are given in percent). In addition, the real contribution of the tracer took into account the underlying carbon transition into the target molecule as described in Chapter 3.5.3.

$$SFL_{corr} = \frac{SFL - SFL_{nat}}{SFL_{Tr} - SFL_{Tr,nat}} \cdot 100 \qquad \text{(Eq. 3)}$$

with

SFL$_{corr}$	for natural background and dilution of ^{13}C tracer enrichment through pre-culture corrected SFL of a proteinogenic amino acid, glucose or riboflavin [%]
SFL	SFL of a proteinogenic amino acid, glucose, or riboflavin calculated according to Equation (1) [%]
SFL$_{nat}$	SFL of proteinogenic amino acid, glucose or riboflavin when synthesized on naturally labeled medium [%]
SFL$_{Tr}$	SFL of the respective tracer at the beginning of the cultivation (0 h), when glycine, formate, glutamate, and yeast extract are fully labeled [%]
SFL$_{Tr,nat}$	SFL of the naturally labeled respective tracer [%]

3.5.2 Correction of LC/MS and ^{13}C NMR data of riboflavin

LC/MS derived MIDs for the 377 *m/z* parent ion of riboflavin were corrected for the natural abundance of isotopes (van Winden et al., 2002). In case, the entire molecule was under investigation, e.g. when [^{13}C$_6$] glucose or fully ^{13}C-labeled yeast extract were fed, all seventeen carbon atoms were regarded as carbon backbone. If only a certain part of the molecule was inspected, e.g. the two-carbon unit derived from glycine, only those carbon atoms were defined as carbon backbone, whereas the rest was defined as carbon atoms of the derivatization. SFLs were derived according to Equations (1) and (3) as described above (Chapter 3.5.1).

Positional ^{13}C enrichment data (%^{13}C$_{exp}$) of single carbon atoms, derived from ^{13}C NMR measurements were corrected in a similar manner according to Equation (4).

$$\%^{13}C_{corr} = \frac{\%^{13}C_{exp} - \%^{13}C_{nat}}{SFL_{Tr} - SFL_{Tr,nat}} \cdot 100 \qquad \text{(Eq. 4)}$$

with

%^{13}C$_{corr}$	the real ^{13}C enrichment of the carbon of interest of riboflavin from a given tracer [%]
%^{13}C$_{exp}$	measured ^{13}C enrichment for carbon atom of interest of riboflavin from a given tracer [%]
%^{13}C$_{nat}$	the natural ^{13}C enrichment of the carbon of interest [%]
SFL$_{Tr}$	SFL of the respective tracer at the beginning of the cultivation (0 h), when glycine, formate, glutamate, and yeast extract are fully labeled [%]
SFL$_{Tr,nat}$	SFL of the naturally labeled respective tracer [%]

In the case of formate addition, ^{13}C NMR results of riboflavin were also corrected for decrease in ^{13}C labeling of the tracer ([^{13}C] formate) over time (appendix Figure 42), which was detected by ^{1}H NMR. Riboflavin concentrations for certain time points were interpolated. With these information riboflavin production was divided into intervals. For every interval the concentration of riboflavin synthesized was calculated (Equation 5) as well as the mean SFL of ^{13}C tracer (Equation 6) and the maximum amount of riboflavin, which could be 100 % ^{13}C labeled at the carbon of interest for that interval (Equation 7).

$$[RF]_i = [RF]_{ti} - [RF]_{ti-1} \qquad \text{(Eq. 5)}$$

$$SFL_{Tr,i} = \frac{\frac{1}{P}\left[(SFL_{Tr,ti} - 1.1) + (SFL_{Tr,ti-1} - 1.1)\right]}{2} \qquad \text{(Eq. 6)}$$

$$[RF]_{max,i} = [RF]_i \cdot SFL_{Tr,i} \cdot \frac{1}{100} \qquad \text{(Eq. 7)}$$

For the entire cultivation the maximally possible amount of 100 % ^{13}C enrichment at the carbon of interest was calculated by forming the sum of the intervals (Equation 8), which was then transformed into percentage of the final riboflavin titer in the cultivation (Equation 9) and the true ^{13}C enrichment in the carbon of interest stemming from the respective ^{13}C tracer (Equation 10).

$$[RF]_{max} = \sum_{i=1}^{n} [RF]_{max,i} \qquad \text{(Eq. 8)}$$

$$\%RF_{13C,max} = \frac{[RF]_{max}}{[RF]_{final}} \cdot 100 \qquad \text{(Eq. 9)}$$

$$\%^{13}C_{corr} = \frac{\frac{1}{P}(\%^{13}C_{meas} - 1.1)}{\%RF_{13C,max}} \cdot 100 \qquad \text{(Eq. 10)}$$

with

$[RF]_i$	riboflavin concentration produced during the interval i (appendix Figure 42) [g L^{-1}]
$[RF]_{ti}$	riboflavin concentration at the end of the interval i [g L^{-1}]
$[RF]_{ti-1}$	riboflavin concentration at the end of the interval i-1 [g L^{-1}]
$SFL_{Tr,i}$	mean SFL of tracer compound at interval i [%]
P	purity of tracer (see Table 3) [%]
$SFL_{Tr,ti}$	SFL of tracer at the end of interval i [%]
$SFL_{Tr,ti-1}$	SFL of tracer at the end of the interval i-1 [%]
$[RF]_{max,i}$	maximal riboflavin concentration with 100 % ^{13}C enrichment from respective tracer for interval i [g L^{-1}]
$[RF]_{max}$	maximal overall riboflavin concentration with 100 % ^{13}C enrichment at carbon of interest from respective tracer [g L^{-1}]
$[RF]_{final}$	final riboflavin concentration after 144 h [g L^{-1}]
$\%RF_{13C,max}$	percentage of final riboflavin titer that is maximally ^{13}C enriched [%]
$\%^{13}C_{meas}$	measured ^{13}C enrichment for carbon atom of interest of riboflavin from given tracer [%]
$\%^{13}C_{corr}$	the real ^{13}C enrichment of the carbon of interest of riboflavin from a given tracer [%]

3.5.3 Calculating the contribution of a ^{13}C-labeled tracer to a target molecule

Table 5: Factor for converting the summed fractional labeling (SFL) of a target molecule into the contribution of a ^{13}C-labeled tracer compound taking into account the biosynthesis of the target molecules. Grey circles, ^{13}C-labeled carbon atoms. White circles, naturally labeled carbon atoms. R indicates the ribityl side chain of riboflavin, comprising five of the seventeen carbon atoms. Ala, alanine; Arg, arginine; For, formate; GCS, glycine cleavage system; Glu, glutamate; Gly, glycine; Ser, serine.

^{13}C Tracer	Target molecule	Maximum SFL [%]	Correction factor F [-]
[^{13}C$_2$] Gly	[^{13}C$_2$] Ser/Ala	67	1.5
[^{13}C] For	[^{13}C$_1$] Ser/Ala	33	3.0
[^{13}C$_2$] Gly (no GCS)	[^{13}C$_2$] Riboflavin	12	8.5
[^{13}C$_2$] Gly (with GCS)	[^{13}C$_3$] Riboflavin	18	5.7
[^{13}C] For	[^{13}C$_1$] Riboflavin	6	17
[^{13}C$_3$] Ser	[^{13}C$_3$] Riboflavin	18	5.7
[^{13}C$_1$] Ser	[^{13}C$_1$] Riboflavin	6	17
[^{13}C$_5$] Glu	[^{13}C$_5$] Arg	83	1.2

In order to take into consideration the maximally possible contribution of a tracer molecule into a target compound, the SFL$_{corr}$ was multiplied by a correction factor F (Table 5). As an example, glycine, a two-carbon compound, can maximally contribute two thirds to e.g. the serine carbon skeleton. This takes into account the biosynthetic route of serine from glycine. Thus, its contribution to a certain molecule does not equal the SFL of that molecule, but can be corrected for the underlying carbon transition. In case of serine, a C$_3$ molecule, the SFL would have to be multiplied by the factor 1.5 in order to get the true contribution (Ψ) of glycine to serine (Equation 11).

$$\Psi_{Tr} = SFL_{corr} \cdot F \qquad \text{(Eq. 11)}$$

with

Ψ_{TR} contribution of a given ^{13}C labeled tracer to the target molecule (amino acid or riboflavin) [%]

SFL_{corr} for natural background and decrease of ^{13}C enrichment of tracer corrected SFL for a target molecule (amino acid or riboflavin) [%]

F   correction factor for underlying carbon transition between tracer and target molecule (see Table 5) [-]

3.6 Transmembrane formate flux simulation

The observed change of ^{13}C enrichment of formate in the culture supernatant indicated an exchange between cell interior and the surrounding medium (Wittmann and Heinzle, 2001). The underlying transmembrane flux of the compound was calculated from experimental data using OpenFLUX (Quek et al., 2009). The model created for this purpose comprised different formate pools, which were fed by ^{13}C formate added to the medium (FOR_{FEED}) and the intracellular formate pool from riboflavin biosynthesis (FOR_{P5P}) (Figure 12). A total of five fluxes were calculated using the model. Two metabolite balances could be formulated:

FOR_{INTR} $v_2 + v_4 - v_3 - v_5 = 0$ (Eq. 12)

FOR_{EXTR} $v_1 + v_3 - v_2 - v_6 = 0$ (Eq. 13)

Together with two sets of measured labeling data for the different formate pools (FOR_{FEED}, FOR_{EXTR}) and two sets of estimated labeling data sets (FOR_{P5P}, FOR_{INTR}) this rendered an overdetermined network. Estimation of intracellular ^{13}C formate labeling is described in Chapter 4.6.1. The deviation between the experimental and simulated data was minimized for the calculated fluxes by the flux software and was taken as best estimate. Subsequent Monte-Carlo analysis provided 90 % confidence intervals for the estimated flux parameters (Wittmann and Heinzle, 2002).

Flux reversibility (ζ) of the transmembrane formate flux was defined as ratio of backward flux to net flux in the forward direction (Equation 14) (Wittmann and Heinzle, 2001).

$$\zeta_{v_2/v_3} = \frac{v_3}{v_2 - v_3} \qquad \text{(Eq. 14)}$$

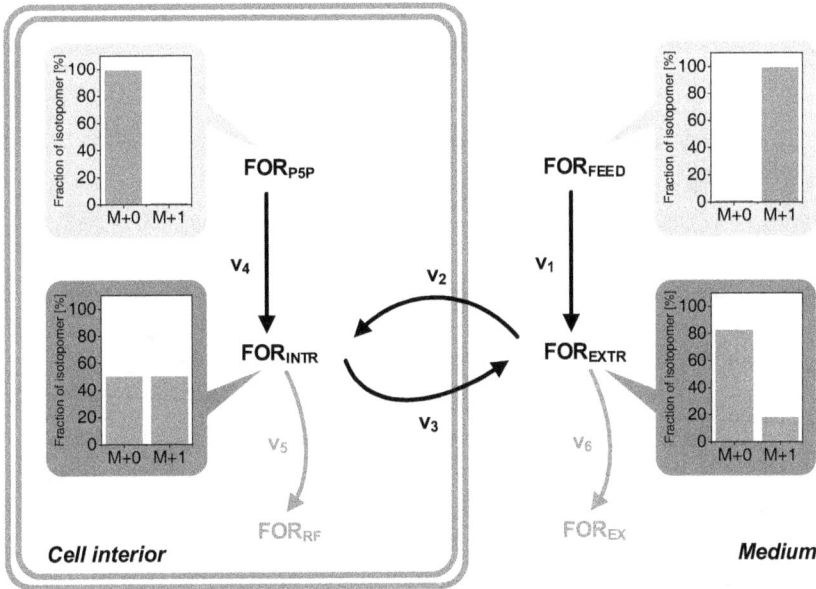

Figure 12: Simplified metabolic model for the simulation of formate fluxes across the plasma membrane. FOR_{FEED} represents the formate pool provided by ^{13}C labeled formate added to the medium, FOR_{EXTR}, formate pool fed by ^{13}C labeled feed formate and intracellular formate pool; FOR_{INTR}, intracellular formate pool, fed by unlabeled formate from pentose 5-phosphate (FOR_{P5P}) and extracellular formate. FOR_{EXTR} was measured and FOR_{INTR} was deduced from extracellular ^{13}C labeling and riboflavin labeling at carbon atom C_2. Those data were used as input for the model (highlighted in red). FOR_{P5P} and FOR_{FEED} were substrates with constant labeling (1.1 % and 99 % ^{13}C enrichment, respectively). The metabolic model is described in the appendix (Chapter 6.9).

41

4 RESULTS AND DISCUSSION

4.1 The benchmark process: growth and riboflavin production on vegetable oil

Industrial riboflavin production with *A. gossypii* is conducted using vegetable oil as carbon source. While other substrates like glucose can also be used for production of the vitamin, vegetable oil is reported to yield the best titers and highest production performance (Demain, 1972). In order to assess growth and productivity of *A. gossypii* WT and B2 used in this study, cultivations were carried out under industrially relevant conditions with complex medium and rapeseed oil as substrate. Furthermore, the medium contained large amounts of yeast extract (29 g L^{-1}) and was supplemented with formate, glycine, and glutamate.

A.gossypii WT and B2 exhibited a two-phase profile (Figure 13): an initial growth phase and a subsequent riboflavin production phase. The overproducer *A. gossypii* B2 achieved a final riboflavin titer of 5.3 g L^{-1} within 144 h (Figure 13A), while the wild type WT showed a titer of 1.5 g L^{-1} at the end of the cultivation (Figure 13D). With a maximum growth rate of 0.07 h^{-1}, a maximum biomass concentration of 20.0 g$_{CDW}$ L^{-1} was reached after 48 h for *A. gossypii* B2. While slight riboflavin production was already initiated at about 24 h, major production started after 36 h when cell growth was almost finished (Figure 13A). During riboflavin biosynthesis, *A. gossypii* B2 did not exhibit growth and cells were rather stationary. After 83 h of cultivation, the strain reached a maximum specific productivity of 17.0 µmol g^{-1} h^{-1} (6.4 mg g^{-1} h^{-1}) (Figure 13C). The different substrates were consumed in a divergent manner. Rapeseed oil was consumed during both phases, but was still available in large amounts at the end of the process. Glycine uptake was rather weak during the initial growth phase, but particularly strong during the riboflavin production phase. However, the observed stoichiometric ratio of 6.1 mol$_{Glycine}$ mol$_{Riboflavin}$$^{-1}$ (Table 6) strongly exceeded the value of 1 mol mol^{-1} expected from the incorporation of one glycine molecule into the pyrimidine ring of the vitamin. Glycine was still present at substantial levels after 144 h (4 g L^{-1}) (Figure 13B). In contrast, the level of glutamate mainly decreased during cell growth, but remained relatively constant towards the end of the process. Within the first 20 h of cultivation, the wild type *A. gossypii* WT reached a final CDW of 9.6 g$_{CDW}$ L^{-1} with a maximum growth rate of 0.17 h^{-1} (Figure 13D). After a transient lag phase, the second growth phase and coupled riboflavin production phase started after 48 h. During this second growth phase, cells grew at a much slower rate of 0.01 h^{-1}, reaching a maximum overall biomass concentration of 28.6 g$_{CDW}$ L^{-1} after 96 h. While cell growth ceased after that, riboflavin production was still ongoing reaching a maximum specific productivity of 3.7 µmol g^{-1} h^{-1} (1.4 mg g^{-1} h^{-1}) after 122 h (Figure 13F). *A. gossypii* WT consumed the substrates rapeseed oil, glycine, and glutamate in a similar manner compared to *A. gossypii* B2: vegetable oil was consumed throughout the cultivation, glutamate consumption was strongest during the first growth phase whereas glycine consumption was greater during riboflavin biosynthesis (Figure 13D, E). Considering the stoichiometric ratio of 250 mol$_{Glycine}$ mol$_{Riboflavin}^{-1}$ (Table 6), glycine uptake and conversion into riboflavin seemed much less efficient for *A. gossypii* WT than B2. The cultivations clearly show that the chosen set-up, i.e. complex medium with vegetable oil, resulted in good growth of the fungus as well as high riboflavin production compared to the riboflavin titers on glucose-containing medium described in the literature (up to about 300 mg L^{-1}) (Ledesma-Amaro et al., 2015c).

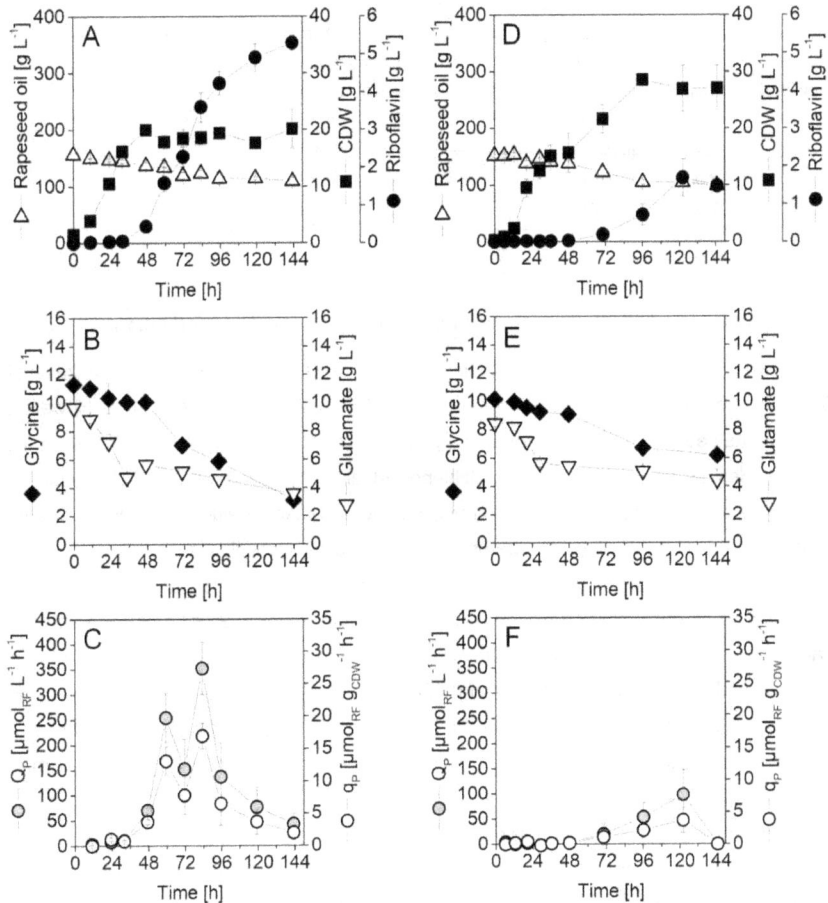

Figure 13: Growth and riboflavin production of *A. gossypii* B2 (A-C) and WT (D-F) on complex medium with rapeseed oil. The data represent mean values and the deviation for three shake flasks, sacrificed per data point. CDW, cell dry weight; Q$_P$, volumetric productivity; q$_P$, specific productivity; RF, riboflavin.

43

Table 6: Growth and production kinetics of *A. gossypii* B2 and WT grown on complex medium with rapeseed oil as main carbon source. Rates (q, Q) and yield coefficients (Y) represent mean values from three independent replicates. RF, riboflavin; Glu, glutamate; Gly, glycine; P, product; S, substrate; X, biomass.

			B2	WT
Rates	$\mu_{max,1}$	$[h^{-1}]$	0.07 ± 0.00	0.17 ± 0.01
	$\mu_{max,2}$	$[h^{-1}]$	-	0.01 ± 0.00
	$q_{S,Oil}$	$[mmol\ g^{-1}\ h^{-1}]$	0.43 ± 0.06	0.75 ± 0.00
	$q_{S,Glu}$	$[mmol\ g^{-1}\ h^{-1}]$	0.14 ± 0.02	0.16 ± 0.00
	$q_{S,Gly}$	$[mmol\ g^{-1}\ h^{-1}]$	0.08 ± 0.02	0.16 ± 0.00
	$q_{P,max}$	$[\mu mol\ g^{-1}\ h^{-1}]$	16.96 ± 1.96	3.69 ± 1.94
	$Q_{P,max}$	$[\mu mol\ L^{-1}\ h^{-1}]$	352.04 ± 52.20	98.60 ± 49.40
Yields	$Y_{X/Oil}$	$[g\ mol^{-1}]$	186.1 ± 21.4	224.1 ± 35.5
	$Y_{X/Glu}$	$[g\ mmol^{-1}]$	0.56 ± 0.08	1.07 ± 0.11
	$Y_{X/Gly}$	$[g\ mmol^{-1}]$	0.98 ± 0.21	1.06 ± 0.12
	$Y_{RF/Oil}$	$[mmol\ mol^{-1}]$	281.6 ± 36.1	183.2 ± 63.8
	$Y_{RF/Glu}$	$[mmol\ mol^{-1}]$	886.4 ± 131.4	7.8 ± 1.1
	$Y_{RF/Gly}$	$[mmol\ mol^{-1}]$	152.4 ± 14.4	4.0 ± 0.7

4.2 Glucose as main carbon source for growth and riboflavin production

Riboflavin production with *A. gossypii* on vegetable oil resulted in final riboflavin titers of $5.3\ g\ L^{-1}$ and $1.5\ g\ L^{-1}$, for *A. gossypii* B2 and WT, respectively (Chapter 4.1). However, glucose has been mainly used to study the pathways for riboflavin production (Bacher et al., 1985; Ledesma-Amaro et al., 2015c; Schlüpen et al., 2003; Silva et al., 2015), given its availability as ^{13}C tracer. First experiments, therefore, aimed to elucidate to which extent glucose enabled to study the microbe under high-productivity conditions, as this would have provided a straightforward strategy for the isotope experiments to be conducted.

The growth and production profiles of *A. gossypii* WT and B2 with $20\ g\ L^{-1}$ glucose as main carbon source are shown in Figure 14. In order to compensate for a drastic decrease in pH the system was buffered with ACES (200 mM) (Good et al., 1966). The two-phase process was characterized by an initial growth and subsequent riboflavin production phase. Riboflavin biosynthesis started after about 12 h: *A. gossypii* B2 accumulated only $74.1\ mg\ L^{-1}$ (0.2 mM) riboflavin after 144 h (Figure 14A). The wild type strain *A. gossypii* WT, on the other hand, produced twice as much riboflavin with a final titer of $155.1\ mg\ L^{-1}$ (0.4 mM) (Figure 14C). This difference in production capacity on glucose is emphasized by volumetric and specific productivities, Q_P and q_P, respectively (Figure 14B and D). *A. gossypii* WT reached the maximum riboflavin productivity of $1.7\ \mu mol\ g^{-1}\ h^{-1}$ after 23 h, right after the growth rate had decreased (Figure 14D, Table 7). *A. gossypii* B2 exhibited its maximum productivity after 35 h with $0.3\ \mu mol\ g^{-1}\ h^{-1}$, which was reduced 5.7-fold compared to the wild type (Figure 14B, Table

7). The fact that the final riboflavin titer only deviated by the factor 2 can be attributed to the production behavior of the two strains: *A. gossypii* WT synthesized about 80 % of the final riboflavin amount within 36 h of the process, while *A. gossypii* B2 continuously produced riboflavin. Interestingly, nearly all the glucose added to the medium was consumed at the end of the exponential growth phase. This left other carbon sources in the medium, like glycine and glutamate (final concentrations 7.2 g L^{-1} and 5.8 g L^{-1} for both strains, respectively) or compounds from the yeast extract for the assembly of riboflavin. *A. gossypii* is known to accumulate lipid bodies inside the cell as storage (Ledesma-Amaro et al., 2014b; Stahmann et al., 1994). It would appear obvious for the cell to mobilize the intracellular triacylglycerides for growth-decoupled riboflavin biosynthesis. Indeed, a previous study showed that the intracellular neutral lipid content increased during glucose consumption and decreased again once glucose had been fully consumed. Monitoring the respiration quotient (RQ) of *A. gossypii* cells, the same study concluded that cells were first growing on glucose, exhibiting an RQ of 1 and once glucose was depleted in the medium, the RQ dropped below 1, indicating growth on a more reduced substrate, i.e. oil (Stahmann et al., 1994).

Both strains exhibited a very similar growth behavior with a maximum growth rate of 0.18 h^{-1} and fairly low biomass yields of 77.1 g mol^{-1} and 67.4 g mol^{-1} for WT and B2, respectively (Table 7). Conversely, the substrate uptake rates were almost the same differing only by a factor 1.1. As far as production and growth behavior are concerned, it is challenging to compare growth rates and productivities to literature data, since most media used differ greatly from the medium applied here. A recently published work compared two different *A. gossypii* strains grown on glucose. The wild type was characterized by a maximum growth rate of 0.04 h^{-1}, a production rate of 0.8 $\mu mol\ g^{-1}\ h^{-1}$ with a final riboflavin titer of about 10 mg L^{-1} (Jeong et al., 2015). However, other studies reported production performances of 30 to 400 mg L^{-1}, depending on the medium and carbon source used (Jiménez et al., 2005; Ledesma-Amaro et al., 2015c; Sugimoto et al., 2009).

The results shown here, clearly demonstrate that riboflavin biosynthesis with *A. gossypii* on glucose does not support overproducing conditions. Riboflavin titers were decreased 9.7-fold and 71.5-fold compared to production on vegetable oil, for *A. gossypii* WT and B2 (Chapter 4.1), respectively. However, the use of glucose presents an interesting starting point in order to gain first insights into the metabolism of the fungus.

Table 7: Growth and production kinetics of *A. gossypii* B2 and WT grown on complex medium with glucose as main carbon source. Rates (q, Q) and yield coefficient (Y) represent mean values from three independent replicates. P, product; S, substrate; X, biomass.

Strain	μ_{max} [h^{-1}]	$Y_{X/S}$ [g mol^{-1}]	q_S [mmol $g^{-1}h^{-1}$]	$q_{P,max}$ [µmol $g^{-1}h^{-1}$]	$Q_{P,max}$ [µmol $L^{-1}h^{-1}$]
B2	0.18 ± 0.01	67.4 ± 2.9	2.6 ± 0.2	0.3 ± 0.1	3.0 ± 0.5
WT	0.18 ± 0.01	77.1 ± 0.6	2.4 ± 0.1	1.7 ± 0.3	15.3 ± 2.4

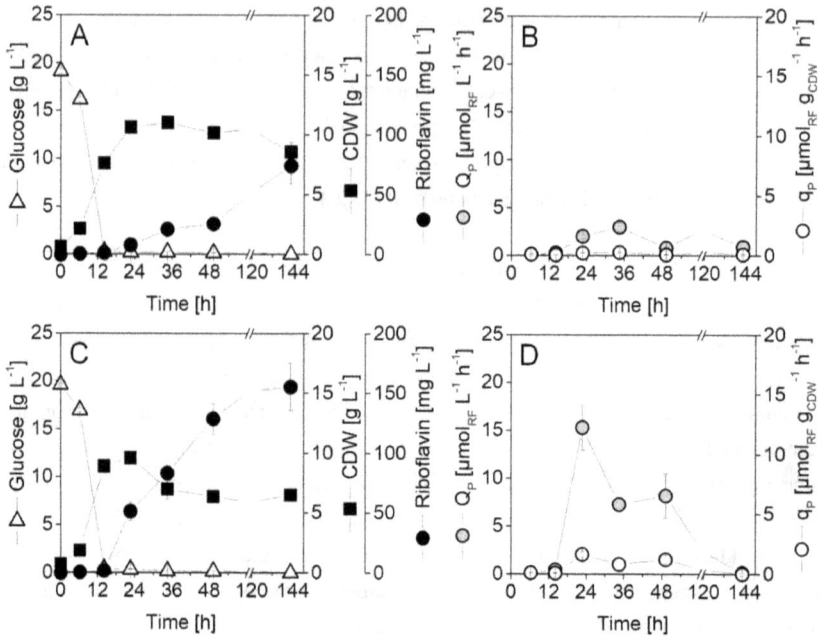

Figure 14: Growth and production profiles of *A. gossypii* B2 (A, B) and WT (C, D). Cells were grown on complex medium with 20 g L^{-1} glucose and 200 mM ACES. The volumetric productivity (Q_P) and specific productivity (q_P) are shown for strain B2 (B) and WT (D). Data were obtained from three independent replicates per time point. CDW, cell dry weight; RF, riboflavin.

4.2.1 First insights into the metabolism with ^{13}C-labeled glucose

In order to get a deeper understanding of the metabolism on glucose, *A. gossypii* WT and B2 cells were grown on [^{13}C$_6$] glucose instead of the naturally labeled sugar (final concentration 20 g L^{-1}). All other medium components remained naturally labeled. This labeling strategy allowed the discrimination of carbon origin of metabolites either from glucose or other medium compounds (Wittmann and Heinzle, 2005). GC/MS measurements of hydrolyzed cell protein, harvested at the end of the exponential growth phase, conveyed valuable insights into growth on glucose. For qualitative inspection of the network, the SFL for each amino acid was calculated from measured MIDs (Table 8, Figure 15). When comparing the two strains investigated, they mostly exhibited a very similar distribution of ^{13}C labeling in their respective metabolism. Most proteinogenic amino acids originated from the naturally labeled portion of the medium, i.e. yeast extract or the supplemented amino acids (Figure 15). However, there were a few exceptions: alanine, glutamate, and aspartate showed strong ^{13}C enrichment with the respective SFL$_{corr}$ ranging from 40 % to 74 % (Table 8). Thus, glucose was not only taken up by the cell, but acted as carbon source during growth despite the complex substrate yeast extract being present in the medium. While glucose transport across the plasma membrane occurs via the energy-dependent PTS in many bacteria (Jahreis et al., 2008), this type of transport has not been proven for yeast (Deutscher et al., 2006). Instead, glucose uptake occurs in an energy-independent manner: proteins that belong to the major facilitator superfamily (MFS) transport hexoses by passive facilitated diffusion along a concentration gradient (Özcan and Johnston, 1999). In *S. cerevisiae* as many as 20 putative hexose transporter genes (*HXT* genes) have been identified, among them transporters as well as sensor proteins (Lin and Li, 2011; Özcan and Johnston, 1999). For *A. gossypii* four putative genes for hexose transporters were identified, representing *HXT*1-7 as well as *HXT*14, and one putative glucose sensor gene was annotated (Lin and Li, 2011). The fact that *S. cerevisiae* contains more transporter genes is suggested to be linked to aerobic fermentation in contrast to complete oxidation of glucose. While *S. cerevisiae* is a so-called Cabtree-positive yeast (De Deken, 1966), *A. gossypii*, on the other hand, prefers to fully oxidize glucose to carbon dioxide through the TCA cycle and the respiratory chain (Merico et al., 2007; Mickelson, 1950). The observed ^{13}C labeling in alanine, 57.3 % and 61.5 % for *A. gossypii* WT and B2, respectively (Table 8, Figure 15) indicated that glucose was taken up by the cell and degraded to pyruvate through glycolysis. Pyruvate was then converted into alanine through the alanine transaminase (Escalera-Fanjul et al., 2017). Alanine was mostly unlabeled (M+0) or fully labeled (M+3) (Table 8), suggesting a straightforward biosynthesis from its metabolic precursor pyruvate as well as uptake from unlabeled yeast extract. Glutamate, even though supplemented to the medium in large amounts, was also heavily ^{13}C-enriched with a SFL$_{corr}$ of 74.1 % and 67.9 % (for WT and B2, respectively). Apparently, the *de novo* formation of glutamate from α-ketoglutarate (AKG), catalyzed by glutamate synthase (Gomes et al., 2014), was strong. The ^{13}C enrichment of glutamate is also an indication for an active TCA cycle. The fact that aspartate also showed ^{13}C enrichment with 48.4 % and 39.9 % for *A. gossypii* WT and B2, respectively, further underlined the active pathway as the metabolic precursor for aspartate is the TCA cycle intermediate oxaloacetate (OAA) (Yagi et al., 1982). The two acidic amino acids exhibited more diverse MIDs compared to alanine. While large fractions were non-labeled (M+0) or fully labeled (M+4 or M+5 for aspartate or glutamate, respectively), isotopomers with both ^{12}C and ^{13}C atoms could also be detected (Table 8). This indicated a molecular rearrangement of fully labeled compounds stemming from glucose and naturally labeled compounds from other medium ingredients of these amino acids through the TCA cycle. The aromatic amino acid tyrosine also showed significant amounts of ^{13}C incorporation (30.6 % and 20.0 % for *A. gossypii* WT and B2, respectively) (Table 8, Figure 15), thus also proving the activity of the PP pathway, since one precursor for aromatic amino acids is erythrose 4-phosphate (E4P) (Braus, 1991). Interestingly, only tyrosine was ^{13}C-enriched and not phenylalanine. This appears noteworthy, since the two amino acids originate from the same pathway (Braus, 1991) and should resemble the same isotopomer distribution. As the medium was based on yeast extract, however, the most likely explanation would be that phenylalanine was not *de novo* synthesized during the growth phase, because it was available in the medium in sufficient amounts whereas the supply of tyrosine was limited.

Glucose was taken up by *A. gossypii* and incorporated into selected amino acids. The ^{13}C labeling of alanine indicated an active glycolysis. Glutamate and aspartate also showed significant amount of ^{13}C enrichment despite large amounts of glutamate in the medium. This proved an active TCA cycle with high conversion of precursor molecules α-ketoglutarate and oxaloacetate into their respective amino acids. Fully labeled proteinogenic tyrosine proved that in addition to the glycolysis also the non-oxidative PP pathway was active.

GLUCOSE YE

Cytosol

PP pathway HIS

Gluconeogenesis

G6P → Ru5P → R5P ──────────→ GAR

Xu5P

Purine biosynthesis

F6P GTP

E4P ← PHE
TYR

S7P

RIBOFLAVIN ──→ RIBOFLAVIN

DHAP ← G3P

C_1 metabolism Riboflavin biosynthesis

THR

3PG → SER ──→ GLY ← GLY

THF CH_2-THF

PEP FOR CHO-THF

ASP ← OAA PYR → AcCoA FOR

PYR

Peroxisome AcCoA ALA
VAL
ILE
LEU

Glyoxylate cycle

ICIT SUC OAA CIT
GLYOX

CIT MAL

AcCoA MAL AKG

GLU ← GLU

TCA cycle

Mitochondrion PRO
ARG

LYS

WT

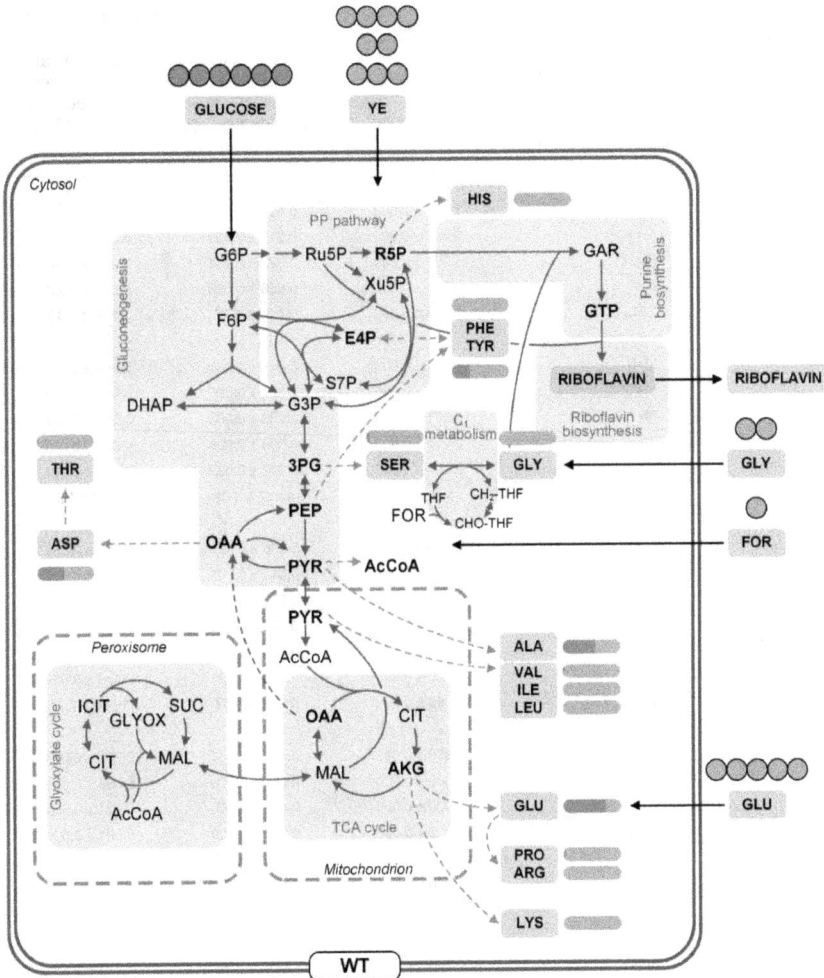

Figure 15: Relative contribution of [$^{13}C_6$] glucose to amino acids from hydrolyzed cell protein of *A. gossypii* WT. Amino acids were obtained at the end of the growth phase of riboflavin producing *A. gossypii* after 12 h. Respective contribution of glucose (red) and other media components (grey) are depicted by the bar next to the respective amino acid. The full length bar represents 100 %. Data represent the corrected summed fractional labeling (SFL$_{corr}$) for the designated amino acids. Data were obtained from three individual replicates. Note that the conversion of citrate to isocitrate via aconitase most likely does not occur in the peroxisome (Murakami and Yoshino, 1997). 3PG, 3-phosphoplycerate; CH_2-THF, 5,10-methylenetetrahydrofolate; AcCoA, acetyl-CoA; AKG, α-ketoglutarate; ALA, alanine; ARG, arginine; ASP, aspartate; CHO-THF, 10-formyltetrahydrofolate; CIT, citrate; DHAP, dihydroxyacetone phosphate; E4P, erythrose 4-phosphate; F6P, fructose 6-phosphate; FOR, formate; G3P, glyceraldehyde 3-phosphate; G6P, glucose 6-phosphate; GAR, glycineamide ribonucleotide; GLU, glutamate; GLY, glycine; GLYOX, glyoxylate; GTP, guanosine triphosphate; HIS, histidine; ICIT, isocitrate; ILE, isoleucine; LEU, leucine; MAL, malate; OAA, oxaloacetate; PEP, phosphoenolpyruvate; PHE, phenylalanine; PP pathway, pentose phosphate pathway; PRO, proline; PYR, pyruvate; R5P, ribose 5-phosphate; Ru5P, ribulose 5-phosphate; S7P, sedoheptulose 7-phosphate; SER, serine; TCA cycle, tricarboxylic acid cycle; THF, tetrahydrofolate; THR, threonine; TYR, tyrosine; VAL, valine; Xu5P, xylulose 5-phosphate; YE, yeast extract.

Table 8: Mass isotopomer distributions (MIDs) of derivatized proteinogenic alanine, aspartate, glutamate, and tyrosine from *A. gossypii* WT and B2 grown on complex medium with [$^{13}C_6$] glucose. Cultivation on naturally labeled glucose served as control. MIDs were corrected for occurrence of natural isotopes. The summed fractional labeling (SFL) for WT and B2 on labeled glucose was corrected as described in Chapter 3.5.1. Note the SFL of the control denotes the calculated SFL of the corrected MID without further correction. For all measured proteinogenic amino acids see appendix Table 16. Data were obtained from three individual replicates.

Analyte		Control	WT	B2
Ala_260	M+0	0.967 ± 0.003	0.402 ± 0.052	0.364 ± 0.073
	M+1	0.032 ± 0.000	0.017 ± 0.002	0.015 ± 0.002
	M+2	0.001 ± 0.000	0.025 ± 0.004	0.022 ± 0.003
	M+3	0.000 ± 0.000	0.556 ± 0.083	0.600 ± 0.054
	SFL$_{corr}$ [%]	**1.13 ± 0.08**	**57.31 ± 5.81**	**61.46 ± 7.15**
Asp_418	M+0	0.962 ± 0.003	0.448 ± 0.045	0.540 ± 0.070
	M+1	0.036 ± 0.000	0.027 ± 0.005	0.028 ± 0.004
	M+2	0.001 ± 0.000	0.036 ± 0.003	0.028 ± 0.002
	M+3	0.000 ± 0.000	0.098 ± 0.008	0.076 ± 0.015
	M+4	0.000 ± 0.000	0.392 ± 0.039	0.327 ± 0.031
	SFL$_{corr}$ [%]	**1.00 ± 0.08**	**48.37 ± 5.08**	**39.92 ± 4.57**
Glu_432	M+0	0.950 ± 0.003	0.136 ± 0.020	0.201 ± 0.020
	M+1	0.049 ± 0.000	0.013 ± 0.001	0.014 ± 0.002
	M+2	0.001 ± 0.000	0.077 ± 0.015	0.078 ± 0.006
	M+3	0.000 ± 0.000	0.082 ± 0.007	0.072 ± 0.012
	M+4	0.000 ± 0.000	0.153 ± 0.038	0.146 ± 0.023
	M+5	0.000 ± 0.000	0.539 ± 0.081	0.489 ± 0.039
	SFL$_{corr}$ [%]	**1.03 ± 0.10**	**74.10 ± 8.37**	**67.90 ± 7.71**
Tyr_466	M+0	0.920 ± 0.007	0.629 ± 0.064	0.731 ± 0.037
	M+1	0.080 ± 0.000	0.054 ± 0.009	0.059 ± 0.009
	M+2	0.000 ± 0.000	0.002 ± 0.000	0.001 ± 0.000
	M+3	0.000 ± 0.000	0.001 ± 0.000	0.001 ± 0.000
	M+4	0.000 ± 0.000	0.001 ± 0.000	0.001 ± 0.000
	M+5	0.000 ± 0.000	0.002 ± 0.000	0.002 ± 0.000
	M+6	0.000 ± 0.000	0.005 ± 0.000	0.003 ± 0.000
	M+7	0.000 ± 0.000	0.007 ± 0.000	0.004 ± 0.000
	M+8	0.000 ± 0.000	0.030 ± 0.005	0.019 ± 0.003
	M+9	0.000 ± 0.000	0.270 ± 0.035	0.178 ± 0.016
	SFL$_{corr}$ [%]	**1.14 ± 0.04**	**31.35 ± 2.97**	**20.03 ± 2.11**

4.2.2 LC/MS measurements reveal first insights into riboflavin metabolism

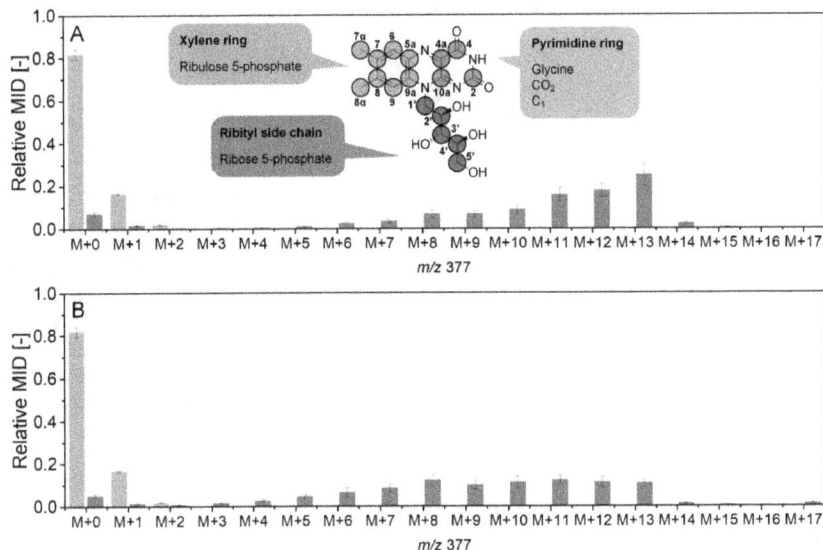

Figure 16: Relative mass isotopomer distributions (MIDs) of riboflavin synthesized by *A. gossypii* WT (A) and B2 (B) grown on complex medium with [$^{13}C_6$] glucose (red). Riboflavin was recovered from the culture broth after 144 h. Riboflavin produced on naturally labeled glucose served as control (grey). The riboflavin structure is depicted (A) with its seventeen carbon atoms, thirteen of which are derived from the pentose 5-phosphate pool while the remaining four have different metabolic origins (potentially glycine, C_1, and carbon dioxide). Data were obtained from three individual replicates.

Riboflavin is a derivative of the heterocyclic compound isoalloxazine and contains seventeen carbon atoms. Its structure is shown in Figure 16A. Pioneering research, starting in the 1950s, aimed to elucidate the molecular origin of all carbon atoms (Bacher et al., 1998; Plaut, 1954a; Plaut, 1954b; Plaut and Broberg, 1956). The vitamin can be divided into three structural parts that differ more or less in their potential carbon origin: the xylene ring, the pyrimidine ring, and the ribityl side chain. While ribulose 5-phosphate and ribose 5-phosphate are the metabolic precursors for the xylene ring and the ribityl side chain, respectively, the remaining four carbon atoms from the pyrimidine ring have a more diverse metabolic origin including the glycine and carbon-one metabolism.

Riboflavin, produced by *A. gossypii* WT and B2 on [$^{13}C_6$] glucose, was recovered from the culture supernatant in small amounts and purified prior to LC/MS analysis. Mass isotopomer distributions from LC/MS spectra are depicted in Figure 16. The SFL$_{corr}$ was calculated from the ^{13}C labeling data: 59 % and 52 % of riboflavin were ^{13}C labeled, for *A. gossypii* WT and B2, respectively (appendix Table 21). This indicated a strong biosynthesis of the vitamin from fully labeled glucose.

The amount of the unlabeled riboflavin isotopomer (M+0) was decreased 12-fold and 16-fold for strains WT and B2, respectively, when compared to riboflavin synthesized on naturally labeled medium. Riboflavin, produced by the two strains, showed a slightly different mass isotopomer distribution: *A. gossypii* WT exhibited a much larger fraction of the M+13 isotopomer than *A. gossypii* B2 (2.3-fold increased). While for *A. gossypii* WT the fraction of isotopomer increased with increasing number of ^{13}C-labeled carbon atoms (Figure 16A), *A. gossypii* B2 showed a more even distribution for mass isotopomers M+7 to M+13 (Figure 16). Interestingly, neither one of the two strains showed significant amounts of isotopomers that had more than thirteen carbon atoms enriched. This left four carbon atoms unlabeled in the riboflavin molecule. These results strongly suggested that the labeled glucose was converted into ribulose 5-phosphate and ribose 5-phosphate and contributed to the thirteen carbon atoms from the ribityl side chain and the xylene ring (Figure 16A). Previous labeling studies using the ^{13}C- or ^{14}C-labeled tracer substrates glucose or ribose, had revealed that the ribityl side chain as well as the xylene ring originate from the same pentose 5-phosphate precursor pool. NMR and radioactivity measurements, which allowed the determination of positional ^{13}C and ^{14}C enrichment, confirmed that these thirteen carbon atoms displayed the same labeling pattern (Bacher et al., 1982; Bacher et al., 1985; Plaut, 1954b; Plaut and Broberg, 1956). In addition, the carbon contribution of the labeled sugars to atoms of the pyrimidine ring seems negligible (Bacher et al., 1982; Bacher et al., 1985). The LC/MS results as shown in Figure 16 match those pioneering results. However, while it can be assumed that the incorporation of labeling into the xylene ring as well as the ribityl side chain occurred, but not into the pyrimidine ring, only ^{13}C NMR measurements would be able to confirm the respective positional enrichments. Unfortunately, the low riboflavin titers generated on glucose did not yield sufficient amounts to enrich and purify riboflavin for ^{13}C NMR analysis.

Riboflavin was more than 50 % ^{13}C-enriched when [^{13}C$_6$] glucose was present in the medium. Thus, glucose clearly had a strong impact on the vitamin. However, this also left about 50 % of the product with unknown carbon origin. Taking into account the reduced production performance of *A. gossypii* on glucose compared to vegetable oil, a different approach had to be chosen for further investigations.

4.3 Replacing oil with its building blocks

The ^{13}C labeling experiments on glucose displayed a starting point in understanding the metabolism of growth and riboflavin production. However, riboflavin production was reduced on glucose and the sugar enters the metabolism through a rather distant pathway compared to vegetable oil, i.e. glycolysis and ß-oxidation, respectively. The two constituents of vegetable oil are glycerol and fatty acids, which are broken down into acetyl-CoA units through ß-oxidation localized in the peroxisome (Vorapreeda et al., 2012). In an attempt to resemble the industrial substrate more closely, acetate and glycerol were added to the medium as major carbon sources, replacing the previous substrate glucose.

4.3.1 Riboflavin production on a mixture of acetate and glycerol

Substrate concentrations were based on the glucose concentration previously used (20 g L^{-1}), with a C-molar ratio of 1:27 for glycerol:acetate (Chapter 3.2.2). In order to counteract pH increase throughout the cultivation MOPS was used as buffer component (250 mM). Growth and production of riboflavin producing A. gossypii WT and B2 are depicted in Figure 17. After 144 h of cultivation, the final riboflavin titer was 8.1 mg L^{-1} and thus only 1.4-fold lower than on glucose for A. gossypii B2 (Figure 17A). The wild type A. gossypii WT reached a final riboflavin titer of 11.9 mg L^{-1} in the culture supernatant at the end of experiment (Figure 17C), which was decreased 13-fold compared to the production on glucose. In contrast to cultivation on glucose, the riboflavin biosynthetic phase already started after 12 h, while growth was still exponential. A. gossypii B2 and WT reached maximum specific productivities of 0.25 µmol g^{-1} h^{-1} (after 24 h) and 0.31 µmol g^{-1} h^{-1} (after 120 h), respectively (Figure 17B and D, Table 9). At a maximum growth rate of 0.09 h^{-1} (Table 9), A. gossypii B2 co-consumed acetate and glycerol reaching a maximum CDW of 3.6 g_{CDW} L^{-1} after 35 h (Figure 17A). At the end of the cultivation, large amounts of acetate (17 g L^{-1}) could still be measured in the culture supernatant. Even though a buffer system was used, the pH increased upon glycerol and acetate uptake (data not shown), which might have prevented further uptake of the substrates and by that inhibited further growth.

Growth of A. gossypii WT on glycerol and acetate proved to be similar to the sister strain B2. There was a slightly longer lag phase at the beginning of the cultivation. Acetate uptake started after 12 h and until then, growth was only minimal on glycerol (Figure 17C). Once, exponential growth had started, the two main carbon sources were consumed in parallel with a specific growth rate of 0.11 h^{-1} (maximum CDW of 2.7 g_{CDW} L^{-1} after 25 h) (Table 9, Figure 17C). While glycerol was almost completely metabolized throughout the cultivation, more than 15 g L^{-1} acetate remained in the medium. Little is known about the uptake systems of the two carbon sources for A. gossypii. In a study, investigating the osmoprotective behavior of A. gossypii, glycerol, which is the main compatible solute for the fungus as well as for S. cerevisiae

(Blomberg and Adler, 1989; Nevoigt and Stahl, 1997), was taken up by a highly active system, which was down-regulated upon hyperosmotic stress (Förster et al., 1998). Most knowledge about transporter systems was gained through experiments with *S. cerevisiae*. The close relative expresses a glycerol/H+ symport (Ferreira et al., 2005). While diffusion through the plasma membrane used to be a generally accepted transport mechanism for glycerol, the polyol is not considered to be able to pass through the lipid bilayer without a transport protein anymore (Klein et al., 2017). The proton symport would also explain the pH increase upon glycerol consumption. Nothing is known about acetate transporters for *A. gossypii*. *S. cerevisiae* contains a permease encoded by the *JEN1* gene (Casal et al., 1999). The proton-symporter also accepts other carboxylic acids such as lactate or pyruvate as substrates. Another gene, *ADY2*, was identified as potential acetate transporter, which was present in cells grown on non-fermentable carbon sources (Casal et al., 1996; Paiva et al., 2004).

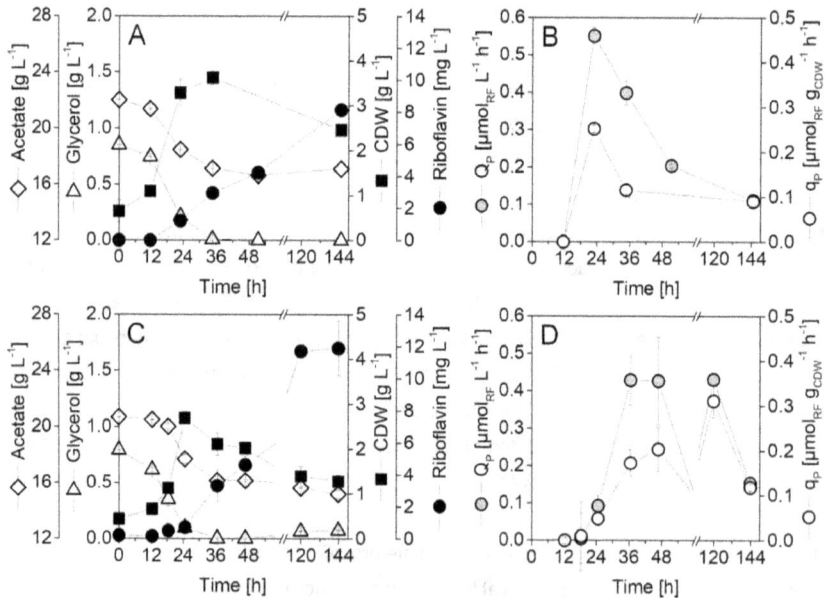

Figure 17: Growth and production profiles of *A. gossypii* B2 (A, B) and WT (C, D). Cells were grown on complex medium with 19.3 g L⁻¹ acetate and 0.7 g L⁻¹ glycerol. The medium was buffered with 250 mM MOPS. The volumetric productivity (Q_P) and specific productivity (q_P) are shown for strain B2 (B) and WT (D). Data were obtained from three independent replicates per time point. CDW, cell dry weight; RF, riboflavin.

Table 9: Growth and production kinetics of *A. gossypii* B2 and WT grown on complex medium with acetate and glycerol as carbon sources. Rates (q, Q) and yield coefficients (Y) represent mean values from three independent replicates. Ac, acetate; Glyc, glycerol; P, product; S, substrate; X, biomass.

			B2	WT
Rates	μ_{max}	[h^{-1}]	0.09 ± 0.01	0.11 ± 0.01
	$q_{S,Ac}$	[mmol g^{-1} h^{-1}]	1.92 ± 0.55	1.82 ± 0.20
	$q_{S,Glyc}$	[mmol g^{-1} h^{-1}]	0.23 ± 0.06	0.80 ± 0.08
	$q_{P,max}$	[μmol g^{-1} h^{-1}]	0.25 ± 0.01	0.31 ± 0.04
	$Q_{P,max}$	[μmol L^{-1} h^{-1}]	0.55 ± 0.02	0.43 ± 0.02
Yields	$Y_{X/Ac}$	[g mol^{-1}]	44.7 ± 4.6	59.7 ± 4.1
	$Y_{X/Glyc}$	[g mol^{-1}]	380.3 ± 41.0	135.8 ± 7.7

4.3.2 Acetate plays a major role in the lower part of the carbon core metabolism

The contribution of the vegetable oil building blocks acetate and glycerol to the metabolism of *A. gossypii* WT was assessed in parallel tracer experiments. In a first set, [$^{13}C_2$] acetate replaced the naturally labeled substance in the medium, while glycerol remained naturally labeled. Most of the amino acids from cell protein were not ^{13}C-labeled (Figure 18, appendix Table 16), indicating a strong uptake of amino acids from yeast extract in the medium. However, ^{13}C incorporation could be detected in single amino acids throughout the metabolism. Acetate is converted into acetyl-coenzyme A by the enzyme acetyl-CoA synthetase (EC 6.2.1.1), which is annotated for *A. gossypii* and was described among others for *S.cerevisiae* and *Schizosaccharomyces pombe* (Martínez-Blanco et al., 1992; van den Berg et al., 1996). It then enters the metabolism either through the TCA cycle located in the mitochondria or the glyoxylate shunt in the peroxisome. Mickelson and Schuler (1953) proposed that oxidation of acetate occurred only through the TCA cycle, arguing that α-ketoglutarate stimulated acetate oxidation the strongest and that *A. gossypii* produced citrate, when both oxaloacetate and acetate were present.

The three amino acids that exhibited the largest ^{13}C enrichment were alanine, aspartate, and glutamate (Figure 18 and Figure 19) with the calculated SFL_{corr} of 16.4 %, 41.3 %, and 63.7 %, respectively for *A. gossypii* WT (Table 10). The mass isotopomer distribution of aspartate nicely proves that acetate from the medium underwent condensation to a four-carbon moiety via the glyoxylate shunt giving rise to the M+4 isotopomer of oxaloacetate-derived aspartate (Figure 19B) (Morin et al., 1992). A study with *S. cerevisiae* showed that the yeast cells were able to grow on acetate even when certain genes of the TCA cycle were deleted, highlighting the potential of the glyoxylate cycle as acetate utilizing pathway (Lee et al., 2011). The increased M+2 fragment of aspartate could originate from the condensation of one fully and one naturally labeled acetyl-CoA in the glyoxylate shunt or a more complex interconversion and re-assembly in the metabolism of the metabolic precursor.

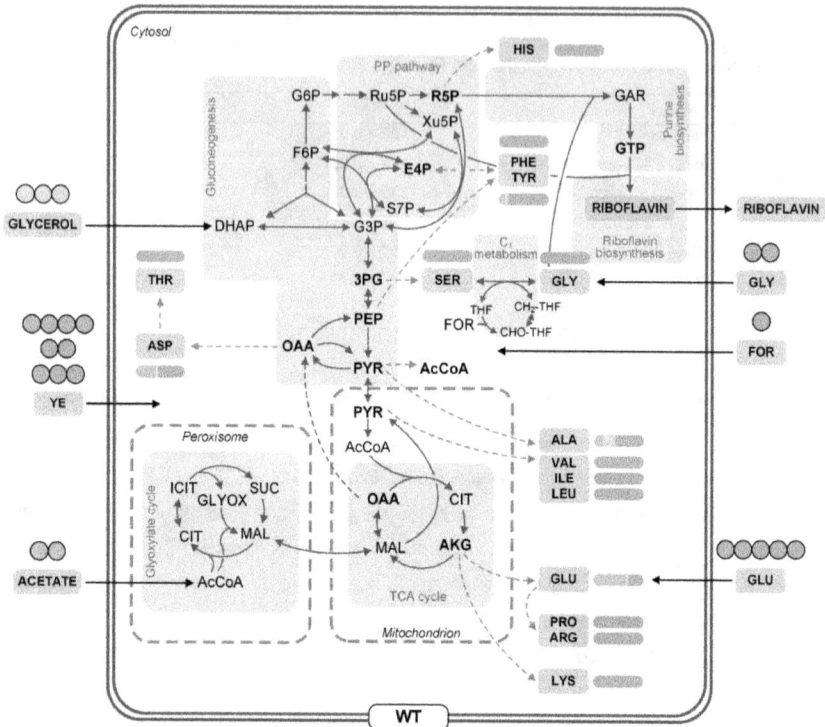

Figure 18: Relative contribution of [$^{13}C_2$] acetate and [$^{13}C_3$] glycerol to amino acids from hydrolyzed cell protein of *A. gossypii* WT grown on complex medium with glycerol and acetate. Respective contribution of acetate (blue), glycerol (green), and other media components (grey) are depicted by the bar next to the respective amino acid. The full length bar represents 100 %. Data represent corrected summed fractional labelings (SFL$_{corr}$) for the designated amino acids. Labeling data were obtained from individual labeling experiments with one of the two carbon sources being fully labeled. Amino acids were obtained at the end of the growth phase of riboflavin producing *A. gossypii* after 24 h. Data were obtained from three individual replicates. Note that the conversion of citrate to isocitrate via aconitase most likely does not occur in the peroxisome (Murakami and Yoshino, 1997). 3PG, 3-phosphoglycerate; CH$_2$-THF, 5,10-methylenetetrahydrofolate; AcCoA, acetyl-CoA; AKG, α-ketoglutarate; ALA, alanine; ARG, arginine; ASP, aspartate; CHO-THF, 10-formyltetrahydrofolate; CIT, citrate; DHAP, dihydroxyacetone phosphate; E4P, erythrose 4-phosphate; F6P, fructose 6-phosphate; FOR, formate; G3P, glyceraldehyde 3-phosphate; G6P, glucose 6-phosphate; GAR, glycineamide ribonucleotide; GLU, glutamate; GLY, glycine; GLYOX, glyoxylate; GTP, guanosine triphosphate; HIS, histidine; ICIT, isocitrate; ILE, isoleucine; LEU, leucine; MAL, malate; OAA, oxaloacetate; PEP, phosphoenolpyruvate; PHE, phenylalanine; PP pathway, pentose phosphate pathway; PRO, proline; PYR, pyruvate; R5P, ribose 5-phosphate; Ru5P, ribulose 5-phosphate; S7P, sedoheptulose 7-phosphate; SER, serine; TCA cycle, tricarboxylic acid cycle; THF, tetrahydrofolate; THR, threonine; TYR, tyrosine; VAL, valine; Xu5P, xylulose 5-phosphate; YE, yeast extract.

GC/MS measurements of glutamate revealed that the M+0 isotopomer was almost 5-fold decreased, indicating a strong *de novo* synthesis from the ^{13}C-labeled acetate (Figure 19C). The remaining 20 % unlabeled isotopomer could have resulted from *de novo* biosynthesis from unlabeled precursor metabolites (Figure 20A) or uptake of naturally labeled glutamate from the medium. Fully labeled proteinogenic glutamate (M+5) was the result of the condensation of fully labeled oxaloacetate and fully labeled acetyl-CoA and subsequent decarboxylation to yield the glutamate precursor α-ketoglutarate (Figure 20F) (Gomes et al., 2014). The increase of M+2 indicated that acetate entered the metabolism at the level of the TCA cycle (Figure 20C) and together with unlabeled oxaloacetate gave rise to [$^{13}C_4$] glutamate.

Acetate had by far the most pronounced impact on building blocks derived from the glyoxylate shunt or TCA cycle. However, substantial amounts of ^{13}C incorporation were also observed for alanine (SFL$_{corr}$ of 16.4 %, Table 10, Figure 19A): in addition to the unlabeled isotopomer (M+0), the M+3 isotopomer was the most abundant, which could be explained by decarboxylation of fully labeled malate to pyruvate, which is the immediate precursor for alanine, through the malic enzyme (Figure 18) (Boles et al., 1998). Alternatively, fully labeled oxaloacetate could be decarboxylated via the PEP carboxykinase to yield phosphoenolpyruvate that in turn could be converted into pyruvate and subsequently to alanine (Zelle et al., 2010).The metabolically more distant amino acid tyrosine was slightly ^{13}C-enriched (SFL$_{corr}$ of 5.5 %), which indicated flux through the PP pathway (Figure 18, Table 10).

Even though the biomass was mainly derived from unlabeled medium compounds, acetate had a strong impact on a few selected amino acids. Acetate played a major role in formation of amino acids derived from the glyoxylate shunt and the TCA cycle. The carbon source, however, also contributed to alanine and the more distant amino acid tyrosine.

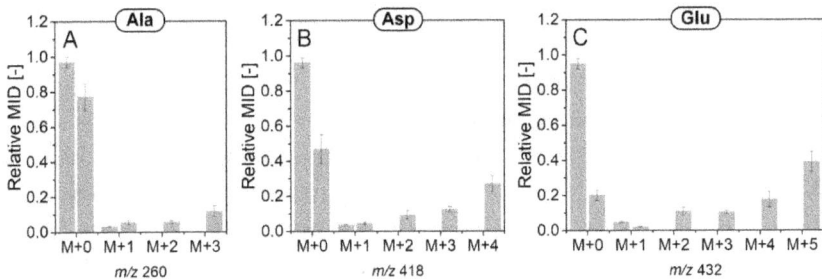

Figure 19: Relative mass isotopomer distributions (MIDs) of proteinogenic alanine (A), aspartate (B), and glutamate (C) from *A. gossypii* WT grown on complex medium with glycerol and acetate as main carbon sources. MIDs are presented for medium with naturally labeled substrates (grey) and [$^{13}C_2$] acetate (blue). Data were obtained from three independent replicates.

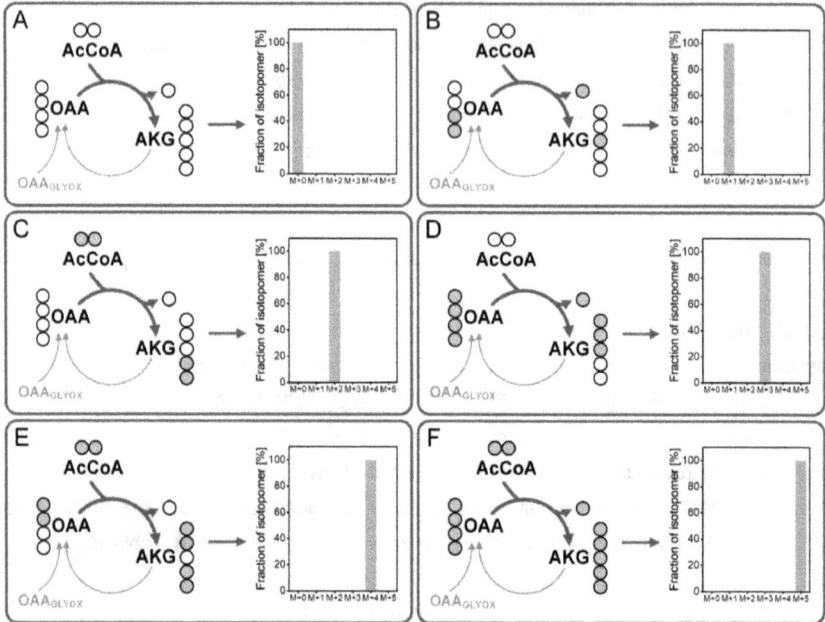

Figure 20: Simplified metabolic model of the tricarboxylic acid (TCA) cycle. Different scenarios are presented for growth on glycerol and fully labeled acetate, which give rise to the different mass isotopomers of α-ketoglutarate (AKG)-derived glutamate. The thin arrows indicate that only the reactions from oxaloacetate (OAA) to AKG are considered, while the origin of OAA can be diverse. White circles, naturally labeled carbon atoms; blue circles, ^{13}C-labeled carbon atoms. AcCoA, acetyl-CoA; GLYOX, glyoxylate shunt.

4.3.3 Glycerol contributes to glycolytic intermediates

In a parallel experiment, $[^{13}C_3]$ glycerol replaced the naturally labeled glycerol, while acetate remained naturally labeled. Several catabolic pathways for glycerol have been described in the literature for fungi (Klein et al., 2017). The dominant pathway for *S. cerevisiae* entails the phosphorylation of glycerol to glycerol 3-phosphate through glycerol kinase (encoded by *GUT1*) and subsequent oxidation to dihydroxyacetone phosphate (DHAP) via the FAD-dependent glycerol 3-phosphate dehydrogenase, encoded by *GUT2* (Klein et al., 2017). Glycerol can also be oxidized to dihydroxyacetone as intermediate by the NAD-dependent glycerol dehydrogenase, which in turn is phosphorylated to DHAP by dihydroxyacetone kinase (encoded by *DAK1/DAK2*) (Klein et al., 2017). The only experimental evidence for the existence of this pathway in yeast was given by a study with *Schizosaccharomyces pombe* (Matsuzawa et al., 2010). A third catabolic pathway involves the intermediate glyceraldehyde, which is then phosphorylated to glyceraldehyde 3-phosphate (G3P) (Klein et al., 2017; Tom et al., 1978). For *A. gossypii*, only the enzymes for the pathway using glycerol 3-phosphate as intermediate are fully annotated in KEGG (Kanehisa et al., 2017; Kanehisa and Goto, 2000; Kanehisa et al., 2016). Due to its close relationship with *S. cerevisiae*, it is likely that this is also the preferred route of glycerol catabolism in the hemiascomycete. However, a whole-genome reannotation study of *A. gossypii* in 2014 revealed two putative genes for glycerol dehydrogenases (Gomes et al., 2014), thereby completing the set of genes necessary for the second catabolic route described above. Regardless of the enzymatic steps chosen for glycerol utilization, glycerol enters the core carbon metabolism at the level of DHAP or G3P. GC/MS measurement of proteinogenic amino acids at the end of the exponential growth phase assessed the fate of glycerol and its metabolic intermediate DHAP and revealed a small but significant contribution of the substrate glycerol to the synthesis of cellular protein (Figure 18, Figure 21, Table 10). The pyruvate-derived amino acid alanine showed a SFL_{corr} of 24.4 % for *A. gossypii* WT (Table 10), indicating a significant contribution of glycerol to its formation. While the non-labeled isotopomer was decreased (Figure 21A), fully ^{13}C-labeled alanine was enriched and accounted for 23 % of the intracellular pool. Incorporation of the ^{13}C labeling from glycerol into alanine indicated an active glycolytic pathway (Figure 21D). Another proof for DHAP oxidation in the glycolytic direction was the observed ^{13}C enrichment of tyrosine (SFL_{corr} of 13.0 %, Table 10). The building blocks for this aromatic amino acid are the glycolytic intermediate phosphoenolpyruvate, which is the immediate precursor of pyruvate, and erythrose 4-phosphate (Braus, 1991). With 4 % fully labeled tyrosine (appendix Table 16), the labeling pattern of this proteinogenic amino acid not only confirmed active glycolysis, but also proved carbon flux through the non-oxidative PP pathway with erythrose 4-phosphate as its intermediate.

Two other proteinogenic amino acids exhibited significant ^{13}C incorporation from glycerol: glutamate and aspartate (SFL_{corr} of 5.6 % and 4.6 % for *A. gossypii* WT, respectively) (Figure 18, Figure 21, Table 10). While the SFL renders information on the carbon origin of a certain molecule, the measured MIDs yield more detailed information on the underlying pathway for a specific metabolite. Glycerol enters the glycolysis, which renders fully labeled pyruvate proven by a fully labeled alanine pool (Figure 21A, D). Pyruvate is then decarboxylated into acetyl-CoA, which enters the TCA cycle (Gey et al., 2008). Thus, the expected mass isotopomer of glutamate for growth on fully labeled glycerol would be the M+2 fragment (Figure 21C, D). Indeed, this fragment is slightly enriched for the proteinogenic amino acid. The second observed isotopomer was M+1, which might have occurred through condensation of non-labeled acetyl-CoA and partially labeled oxaloacetate. Apart from the TCA cycle, oxaloacetate can originate from various metabolic reactions: carboxylation of pyruvate, transamination of glutamate from the medium (Wood, 1968), assimilation of acetate from the glyoxylate shunt (Lee et al., 2011). The observed MID for proteinogenic aspartate revealed that the M+1 isotopomer was the largest ^{13}C-labeled fraction, which indicated glycerol conversion via glycolysis and subsequent TCA cycle (Figure 21B, D).

The isotope experiment with [^{13}C$_3$] glycerol and unlabeled acetate highlighted the intracellular fate of the polyol. While it contributed to the formation of amino acids derived from the TCA cycle, glycerol had the strongest impact on glycolytic intermediates. This was derived from the ^{13}C labeling of the amino acids alanine and tyrosine: both exhibited to some extent their fully labeled isotopomer.

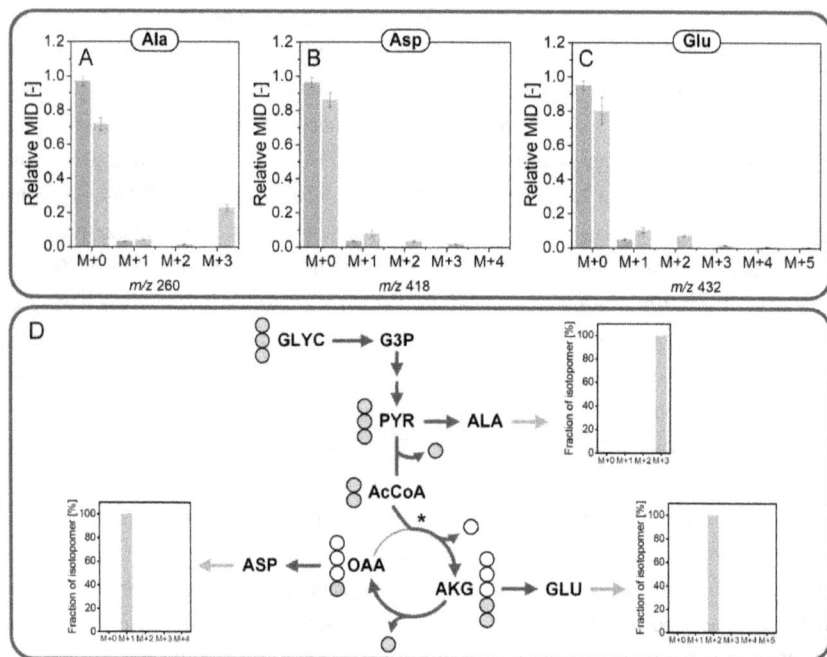

Figure 21: Relative mass isotopomer distributions (MIDs) of proteinogenic alanine (A), aspartate (B), and glutamate (C) from *A. gossypii* WT grown on complex medium with glycerol and acetate as main carbon sources. MIDs are presented for medium with naturally labeled substrates (grey) and [^{13}C$_3$] glycerol (green). (D) Depicts a simplified metabolic model for the lower glycolysis and the tricarboxylic acid (TCA) cycle, which highlights the fate of fully labeled glycerol in the metabolism of *A. gossypii*. The labeling of the intermediates pyruvate, oxaloacetate (OAA), and α-ketoglutarate (AKG) were derived from the amino acids alanine, aspartate, and glutamate, respectively. The asterisk indicates that here, only the condensation of fully labeled acetyl-CoA (AcCoA) with naturally labeled OAA was considered, which was done for reasons of simplicity. White circles, naturally labeled carbon atoms; green circles, ^{13}C-labeled carbon atoms. G3P, glyceraldehyde 3-phosphate, GLYC, glycerol; PYR, pyruvate.

Table 10: Summed fractional labeling (SFL) of derivatized amino acids from hydrolyzed cell protein of riboflavin producing *A. gossypii* WT and B2, grown on complex medium with glycerol and acetate as main carbon sources with different ^{13}C tracers. The SFL was determined using the original measurement data (appendix Table 16) (left column), and was corrected for the naturally occurring isotopes in the carbon backbone as well as purity of the tracer (SFLcorr, right column). The correction is described in Chapter 3.5.1. Data were obtained from three individual replicates.

	Control	WT				B2			
		$[^{13}C_3]$ Glycerol		$[^{13}C_2]$ Acetate		$[^{13}C_3]$ Glycerol		$[^{13}C_2]$ Acetate	
	SFL	SFL	SFLcorr	SFL	SFLcorr	SFL	SFLcorr	SFL	SFLcorr
	[%]	[%]	[%]	[%]	[%]	[%]	[%]	[%]	[%]
Alanine	1.13 ± 0.07	25.26 ± 2.71	24.44 ± 2.62	17.31 ± 1.44	16.40 ± 1.36	12.77 ± 1.30	11.82 ± 1.20	19.42 ± 2.02	18.53 ± 1.93
Aspartate	1.00 ± 0.08	5.62 ± 0.72	4.59 ± 0.59	41.91 ± 4.53	41.26 ± 4.46	2.53 ± 0.41	1.47 ± 0.24	39.75 ± 4.54	39.07 ± 4.46
Glutamate	1.03 ± 0.03	6.56 ± 0.60	5.55 ± 0.51	64.08 ± 5.71	63.65 ± 5.67	2.95 ± 0.22	1.90 ± 0.14	66.63 ± 7.00	66.22 ± 6.95
Tyrosine	1.14 ± 0.08	13.93 ± 1.73	12.99 ± 1.61	6.54 ± 0.70	5.53 ± 0.59	5.56 ± 0.82	4.54 ± 0.67	4.65 ± 0.59	3.61 ± 0.46

4.3.4 *A. gossypii* B2 has a reduced glycolytic flux on glycerol and acetate

Cultivations with fully labeled glycerol and acetate were also carried out with *A. gossypii* B2 (Table 10, appendix Table 16). Since the observed labeling patterns showed high similarity between the two strains, they will not be discussed in great detail. Nevertheless, it is important to note that while the impact of acetate on the metabolism was almost unchanged for WT and B2, glycerol seemed to be metabolized slightly differently. In *A. gossypii* WT, the polyol contributed one fourth to the intracellular alanine and 13 % to the intracellular tyrosine pool (Figure 18, Figure 21, Table 10). Both amino acids stem from the glycolytic conversion of glycerol to intermediates like phosphoenolpyruvate and pyruvate. When *A. gossypii* B2 was grown on fully labeled glycerol, ^{13}C incorporation of alanine was decreased by the factor 2 (SFL of 12.8 % compared to 25.3 % for *A. gossypii* WT) (Table 10). The same effect was observed for tyrosine (SFL of 5.6 % instead of 13.9 %). These data strongly indicated a reduced carbon flux from glycerol to the glycolytic intermediates phosphoenolpyruvate and pyruvate. This might suggest an increased utilization of carbon through gluconeogenesis, which would mirror growth and production physiology of *A. gossypii* grown on vegetable oil, the carbon source of choice for this process. Nevertheless, these data suggest that while there are small differences between the two strains, *A. gossypii* WT and B2 metabolize the alternative carbon sources acetate as well as glycerol in a very similar manner.

4.4   Characterization of growth physiology on vegetable oil via GC/MS analysis

The alternative carbon sources glucose and the building blocks of oil, glycerol and acetate, resulted in drastically decreased production performance compared to vegetable oil. For *A. gossypii* B2, the final riboflavin titer was decreased 72-fold and 675-fold for glucose and acetate:glycerol, respectively. This highlighted the enormous impact of the carbon source on product formation and clearly showed that the investigation of the fungus under high riboflavin production conditions required vegetable oil as main carbon source. Consequently, for further investigations regarding growth and the subsequent riboflavin biosynthetic phase, complex medium with rapeseed oil was used. A first set of ^{13}C tracer experiments aimed at the elucidation of metabolic processes during the initial growth phase. Since *A. gossypii* WT and B2 showed almost identical labeling patterns, only the results for the overproducing strain B2 are discussed.

4.4.1   Metabolic origin of amino acids reveals interconversion of riboflavin precursors

The addition of $[^{13}C_2]$ glycine, a known riboflavin building block, should investigate the potential withdrawal of this amino acid into other metabolic pathways. Glycine in the cell protein of *A. gossypii*, analyzed at the end of the growth phase, revealed substantial ^{13}C enrichment and identified the added glycine as dominant source for the intracellular pool (Figure 22A, Figure 23). During the same period, the ^{13}C enrichment of glycine in the cultivation broth decreased from 97 % in the initial medium to 88 % at the point in time of harvesting (appendix Figure 40), indicating a weak but significant formation of the amino acid from other sources and exchange across the membrane. In order to account for dilution of labeling due to naturally labeled pre-culture medium, in a parallel experiment glycine, formate, glutamate, and yeast extract were fully labeled. The ^{13}C enrichment of the culture supernatant at the beginning of the cultivation was used to normalize the data from respective tracer experiments (appendix Figure 39). On the basis of these data, exogenous glycine obviously provided 85 % of the intracellular pool (Table 11). The use of $[^{13}C_2]$ glycine further resulted in elevated ^{13}C enrichment of proteinogenic serine (Figure 22B, Figure 23). The major fraction of serine was double labeled, i.e. $[^{13}C_2]$ serine, indicating incorporation of the entire glycine carbon skeleton. On the basis of the SFL, about 63 % of serine stemmed from exogenous glycine (Table 11). This calculation considered the fact that 100 % enriched glycine provides 67 % enriched serine, as only two labeled carbons contribute to the three carbon skeleton of serine (Table 5). The observed SFL for serine of 42 % was corrected accordingly by a factor of 1.5. One should note that about 1 g L^{-1} serine was initially present in the medium as part of the yeast extract, potentially having an impact on the process.

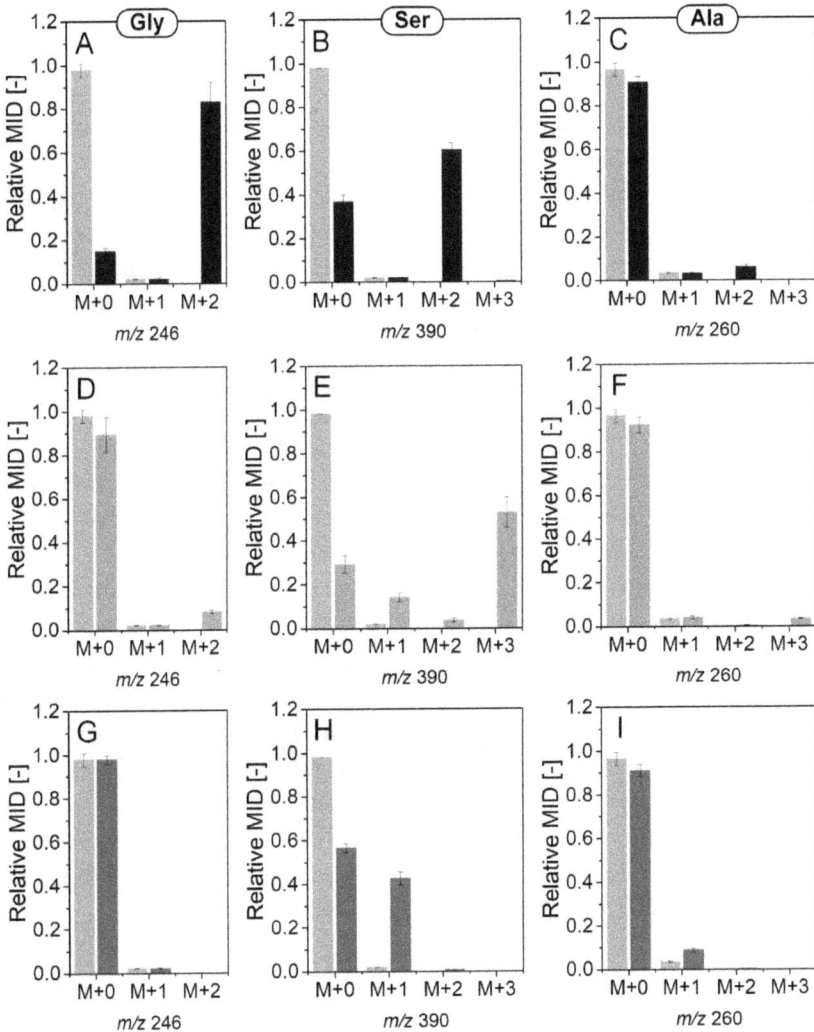

Figure 22: Relative mass isotopomer distributions (MIDs) of proteinogenic glycine (A, D, G), serine (B, E, H), and alanine (C, F, I) from *A. gossypii* B2 grown on complex medium with rapeseed oil as main carbon source. MIDs are presented for medium with naturally labeled substrates (grey), [$^{13}C_2$] glycine (blue), [$^{13}C_3$] serine (green), and [^{13}C] formate (red). Data were obtained from three individual replicates.

In order to study the obvious connection between glycine and serine, [$^{13}C_3$] serine was now added to the initial medium in a next experiment. It was interesting to note that intracellular serine did not only reflect the labeling pattern of the used serine tracer, but also contained significant amounts of single labeled, i.e. $[^{13}C_1]$ serine (Figure 22E), resulting from the reversible nature of serine hydroxymethyltransferase: the triple labeled serine was cleaved into double labeled glycine and a single labeled carbon-one unit. Both labeled products were then re-assembled into serine with non-labeled counterparts, providing the observed single and double labeled serine mass isotopomers. Overall, the externally present serine strongly supplied the intracellular serine pool (Figure 22E). The SFL_{corr} was 63 % (Table 11). The remaining 37 % of the serine pool originated from other sources. Obviously the impact of external glycine on the intracellular serine pool, i.e. 63 % versus maximally 37 %, was diminished when serine was supplemented in the tracer study. Exogenous serine further contributed 9 % to the intracellular glycine pool. Taking both experiments together, the interconversion was obviously highly reversible, indicating a tight metabolic connection between the two amino acids.

A third tracer study investigated the metabolic fate of exogenous formate. When using $[^{13}C]$ formate, proteinogenic serine was strongly labeled and exhibited a SFL_{corr} of 14 % (Figure 22H, Figure 23, Table 11). In order to derive the corresponding relative flux from formate to serine, again the dilution effect had to be considered i.e. fully ^{13}C-labeled formate donates only one carbon atom to serine, leading to a maximum of 33 % enrichment. As a consequence, the cells derived 42 % of their intracellular serine from external formate. The observed fraction of non-labeled intracellular serine (Figure 22H) was an indication that A. gossypii supplied the carbon-one units also from other substrates, involving also serine from the added yeast extract. A wider analysis of the ^{13}C labeling patterns of amino acids from the cell protein revealed that all three tracers contributed to the metabolically more distant amino acid alanine (4 %, 4 %, and 2 % from glycine, serine, and formate, respectively) (Figure 22C, F, I and Figure 23,). In contrast, the other amino acids were not enriched in ^{13}C in any of the experiments (appendix Table 17).

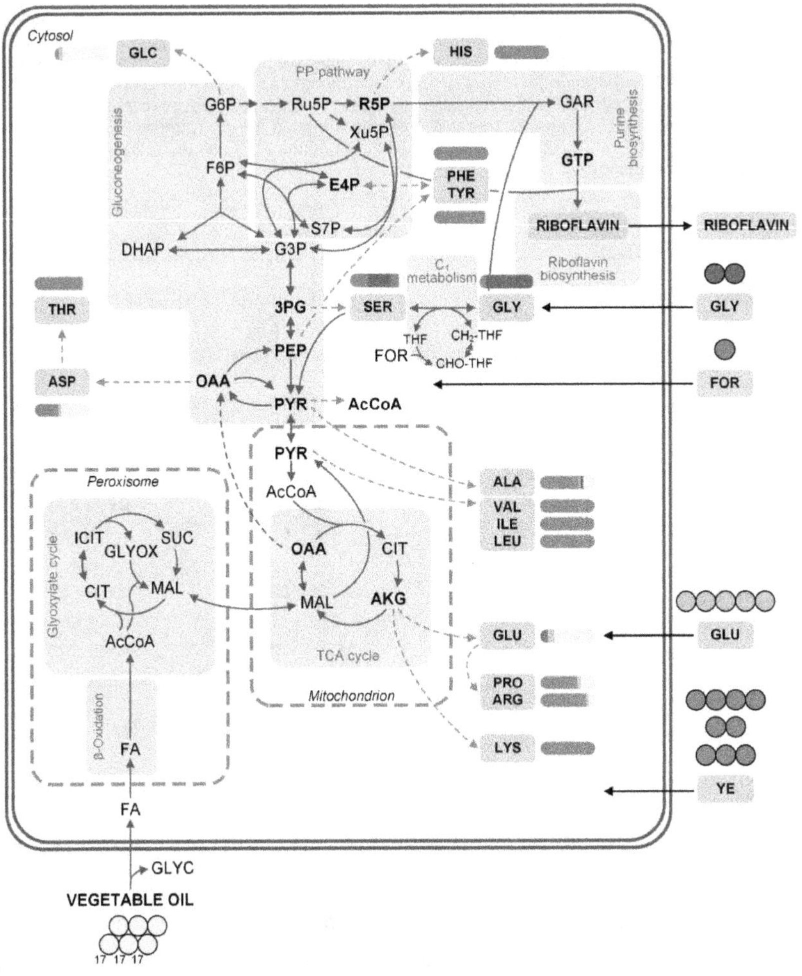

Figure 23: Relative contribution of ^{13}C tracers to amino acids from hydrolyzed cell protein and glycogen of *A. gossypii* B2 grown on complex medium with rapeseed oil. Metabolites were obtained at the end of the growth phase of riboflavin producing *A. gossypii* after 36 h. Respective contribution of $[U^{13}C]$ yeast extract (purple), $[^{13}C_5]$ glutamate (light blue), $[^{13}C_2]$ glycine (blue), $[^{13}C]$ formate (red), and other media components (grey) are depicted by the bar next to the respective amino acid. The full length bar represents 100 %. Data represent summed fractional labelings (SFL$_{corr}$), corrected for natural labeling and dilution effects through unlabeled pre-culture medium. Note that the conversion of citrate to isocitrate via aconitase most likely does not occur in the peroxisome (Murakami and Yoshino, 1997). Data were obtained from three independent replicates. 3PG, 3-phosphoplycerate; CH$_2$-THF, 5,10-methylenetetrahydrofolate; AcCoA, acetyl-CoA; AKG, α-ketoglutarate; ALA, alanine; ARG, arginine; ASP, aspartate; CH$_2$-THF, 5,10-Methylenetetrahydrofolate; CHO-THF, 10-formyltetrahydrofolate; CIT, citrate; DHAP, dihydroxyacetone phosphate; E4P, erythrose 4-phosphate; F6P, fructose 6-phosphate; FA, fatty acids; FOR, formate; G3P, glyceraldehyde 3-phosphate; GAR, glycineamide ribonucleotide; G6P, glucose 6-phosphate; GLU, glutamate; GLY, glycine; GLYC, glycerol; GLYOX, glyoxylate; GTP, guanosine triphosphate; HIS, histidine; ICIT, isocitrate; ILE, isoleucine; LEU, leucine; MAL, malate; OAA, oxaloacetate; PEP, phosphoenolpyruvate; PHE, phenylalanine; PP pathway, pentose phosphate pathway; PRO, proline; PYR, pyruvate; R5P, ribose 5-phosphate; Ru5P, ribulose 5-phosphate; S7P, sedoheptulose 7-phosphate; SER, serine; TCA cycle, tricarboxylic acid cycle; THF, tetrahydrofolate; THR, threonine; TYR, tyrosine; VAL, valine; Xu5P, xylulose 5-phosphate; YE, yeast extract.

65

Table 11: Summed fractional labeling (SFL) of derivatized amino acids from hydrolyzed protein of riboflavin producing *A. gossypii* B2, grown on complex medium with rapeseed oil as main carbon sources with different ^{13}C tracers. The SFL was determined using the original measurement data (appendix Table 17) (left column), and was corrected for the naturally occurring isotopes in the carbon backbone as well as tracer dilution through naturally labeled pre-culture medium (center column). The right column displays the values that also take into consideration the biosynthetic route and carbon transition of the molecules. The correction is described in Chapters 3.5.1 and 3.5.3. Data were obtained from three individual replicates.

	SFL [%]	SFLcorr [%]	Contribution Ψ [%]
[$^{13}C_2$] Glycine			
Alanine	5.20 ± 0.11	4.19 ± 0.09	6.29 ± 0.13
Glycine	84.10 ± 10.64	85.38 ± 10.80	85.38 ± 10.80
Serine	41.59 ± 2.29	41.67 ± 2.29	62.51 ± 3.44
[^{13}C] Formate			
Alanine	3.01 ± 0.11	2.00 ± 0.07	6.00 ± 0.22
Glycine	1.25 ± 0.16	0.00 ± 0.00	0.00 ± 0.00
Serine	14.70 ± 0.85	14.23 ± 0.82	42.69 ± 2.47
[$^{13}C_3$] Serine			
Alanine	4.89 ± 0.11	4.05 ± 0.09	4.05 ± 0.09
Glycine	9.44 ± 1.13	8.91 ± 1.07	8.91 ± 1.07
Serine	60.13 ± 3.61	63.14 ± 3.79	63.14 ± 3.79
Control			
Alanine	1.15 ± 0.08	0.05 ± 0.08	
Glycine	1.15 ± 0.08	0.05 ± 0.08	
Serine	1.05 ± 0.11	-0.05 ± 0.11	

4.4.2 Glycine is efficiently taken up to fuel the intracellular pool

Glycine is a key player of riboflavin production in *A. gossypii* as reported before for different strains and cultivation conditions (Demain, 1972). Therefore, important questions revolve around the pathways that supply this important precursor. Intracellular glycine is accessible from a variety of different reactions (Figure 24A). It can be (i) either taken up directly from the medium or (ii) originate from serine and threonine involving SHMT (Schlüpen et al., 2003) and threonine aldolase (Monschau et al., 1998), respectively. In contrast, glyoxylate aminotransferase potentially also forming glycine is absent in *A. gossypii* (Gomes et al., 2014; Kato and Park, 2006). The conducted tracer studies allowed to distinguish between the forming routes. During the growth phase, 85 % of intracellular glycine were taken up from extracellular [$^{13}C_2$] glycine. In a similar way, the related yeast *S. cerevisiae* also exhibits a strongly regulated biosynthesis. Only upon starvation of a given amino acid its biosynthetic pathway is activated (Ljungdahl and Daignan-Fornier, 2012). Amino acid transport systems have not been studied in *A. gossypii*. The yeast *S. cerevisiae* expresses a general amino acid permease (*GAP1*) for the unspecific glycine transport across the membrane (Cain and Kaiser, 2011; Wipf et al., 2002). The other routes to intracellular glycine played only a minor role. This also included threonine aldolase. In previous work, overexpression of the corresponding *GLY1* gene along with threonine supplementation resulted in 9-fold increased riboflavin productivity (Monschau et al., 1998), admittedly at much lower production performance.

Once serine was present in the medium, the cells switched to uptake from the medium. Under these conditions, intracellular serine was largely taken up (63 %), with the remaining fraction being supplied from glycine (Figure 24B). In cases, where serine was not added, i.e. the standard production set-up, glycine supplied even larger amounts of serine (63 %) and seemed to function as an efficient precursor (Table 11). Glycine and serine were closely linked via the highly reversible SHMT. The extent to which serine or glycine were labeled from the respective tracer strongly depended on substrate availability (Figure 22, Figure 23). The enzyme adjusted the pools of glycine and serine in a highly flexible manner, depending on the nutrient status. It seems that the enzyme has a high balancing capacity, which is probably required to support the supply of both amino acids under the highly dynamic process conditions. Threonine, another amino acid participating in the glycine metabolism of *A. gossypii*, was not labeled, whether ^{13}C serine or ^{13}C glycine were fed (appendix Table 17). This matches the irreversibility of threonine aldolase, which catalyzes the unidirectional conversion of threonine to glycine (Schlüpen et al., 2003).

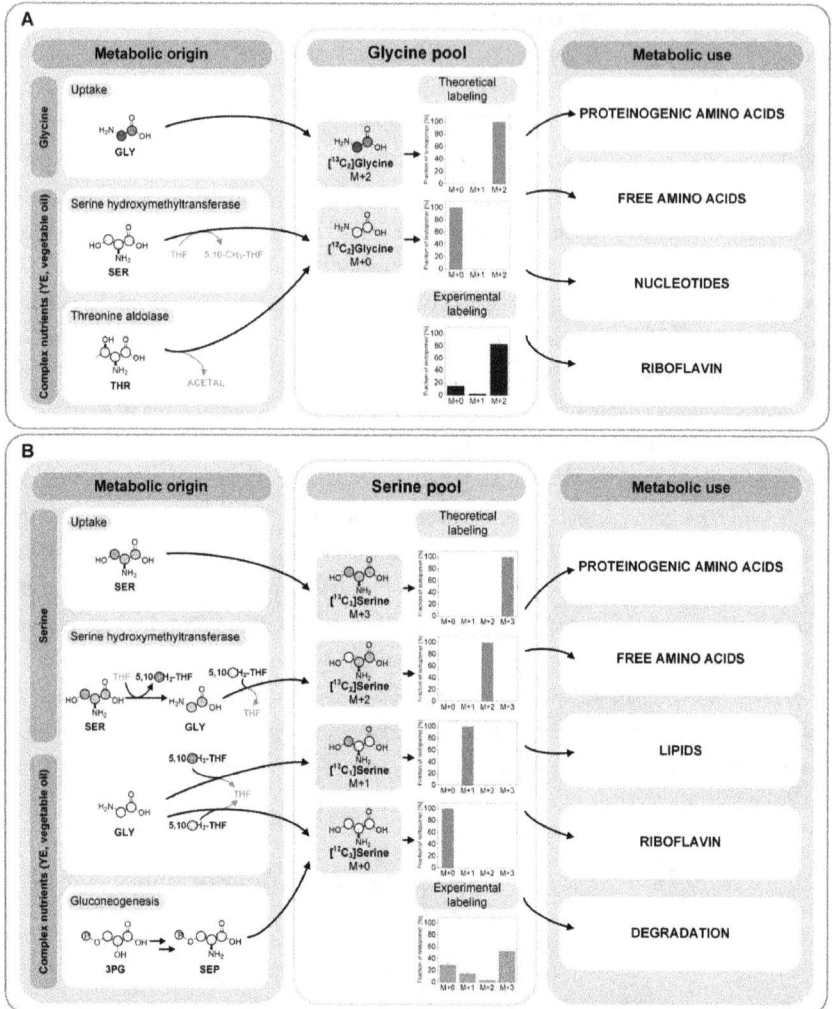

Figure 24: Contribution of different metabolic routes to glycine (A) and serine (B) in *A. gossypii* B2. The expected theoretical mass isotopomer distribution for the resulting pools, inferred from pathway stoichiometry and related carbon transition, are indicated as well as the actually measured labeling of the intracellular pool. Data were obtained from three individual replicates. Colored circles, ^{13}C-labeled carbon atoms; white circles, naturally labeled carbon; 3PG, 3-phosphoglycerate; 5,10-CH$_2$-THF, 5,10-methylenetetrahydrofolate; ACETAL, acetaldehyde; GLY, glycine; SER, serine; SEP, 3-phosphoserine; THF, tetrahydrofolate; THR, threonine.

4.4.3   Serine degradation pathways cause undesired loss of riboflavin precursors

In tracer studies with ^{13}C serine, ^{13}C glycine, and also with ^{13}C formate, the pyruvate derived amino acid alanine was significantly enriched in ^{13}C (Figure 22). Metabolically, serine, glycine, and formate are connected via SHMT potentially allowing for an exchange of the ^{13}C labeling between the pools. Indeed, glycine and (indirectly) formate were converted into serine (Figure 23). Since each of the different tracers donated the label to alanine, serine appeared as the connecting node to alanine. One way of serine utilization is reconversion to 3-phosphoglycerate via hydroxypyruvate and glycerate. Potentially, 3-phosphoglycerate could indeed serve as precursor for pyruvate and then alanine through the action of enzymes of the lower glycolytic chain. However, growth on vegetable oil in the absence of glycolytic substrate rather relies on gluconeogenesis, operating the pathway in the opposite direction. Moreover, one might expect ^{13}C labeled aromatic amino acids derived from ^{13}C glycolytic intermediates, which were not found (Figure 23, appendix Table 17), suggesting that such a glycolytic route is probably not involved. Indeed, this pathway is only found in methylotrophic microbes as part of the serine cycle (Chistoserdova and Lidstrom, 1994) and in mammals and plants (Liepman and Olsen, 2001; Liepman and Olsen, 2003; Snell, 1984), but has apparently not been described in fungi or yeasts. Admittedly, this is not a clear disproval, because the absence of ^{13}C enrichment in aromatic amino acids could also be explained by the fact that they are taken up from yeast extract, rather than being synthesized *de novo*. Other pathways might lead to alanine formation from serine that could not be identified in this set-up. As shown, serine degradation, together with the reversible action of SHMT, caused continuous withdrawal of carbon from the glycine pool.

4.4.4   The amino acid *de novo* biosynthesis is tightly regulated with a few exceptions

Yeast extract was part of the medium in large amounts (29 g L^{-1}). It was, therefore, of great importance to quantify the contribution of yeast extract to cell growth and production. In a first experiment, the formation of cell building blocks from yeast extract should be investigated. Since fully ^{13}C-labeled yeast extract was not readily available, it was custom-synthesized in cooperation with the company Ohly (Hamburg, Germany). A yeast strain, typically used for the production of yeast extract, was grown on [$^{13}C_6$] glucose under fed-batch conditions. After centrifugation and autolysis of the cells, the biomass was dried by lyophilization. The composition was analyzed and compared to the conventional extract of this yeast strain, whose suitability for riboflavin production with *A. gossypii* had been tested in preliminary studies (data not shown). Since the differences in composition were negligible, the ^{13}C-labeled extract was appropriate for the use in ^{13}C tracer studies with *A. gossypii*. When fully ^{13}C-labeled yeast extract was added to the medium and substituted the naturally labeled yeast extract, proteinogenic amino acids harvested at the end of the exponential growth phase, were largely

^{13}C-labeled (Figure 23, appendix Table 18). This indicated a strong uptake of yeast extract-based components and subsequent incorporation into the cell protein (Figure 23, appendix Table 18). While many amino acids were derived entirely from the complex source, a few selected amino acids exhibited less ^{13}C incorporation: alanine, aspartate, and glutamate (Figure 25). This indicated a different metabolic origin for those amino acids. Serine and glycine showed comparatively low ^{13}C enrichments with SFL$_{corr}$ of 32 % and 12 %, respectively (appendix Table 18). This matched the observation that glycine was identified as major contributor to the intracellular serine and glycine pools (Chapter 4.4.1).

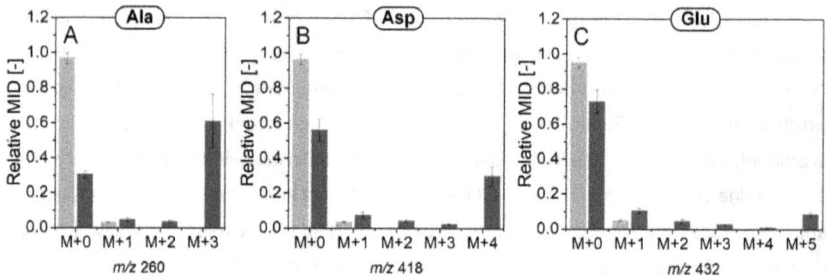

Figure 25: Relative mass isotopomer distributions (MIDs) of proteinogenic alanine (A), aspartate (B), and glutamate (C) from *A. gossypii* B2 grown on complex medium with rapeseed oil as main carbon source. MIDs are presented for medium with naturally labeled substrates (grey) and [U^{13}C] yeast extract (purple). Data were obtained from three individual replicates.

The amino acids aspartate, alanine, and glutamate were less ^{13}C-enriched, when compared to the other amino acids (SFL$_{corr}$ of 43 %, 72 %, and 14 %, respectively) (Figure 23, Figure 25, appendix Table 18). Of the three, proteinogenic alanine exhibited the largest fully labeled isotopomer (M+3) fraction with 61 %. Unlabeled alanine from cell protein made up 31 % of the intracellular pool (Figure 25A). The majority of aspartate was non-labeled, however, as much as 30 % of the pool were fully labeled (M+4), which indicated origin from the fully labeled yeast extract (Figure 25B). The intracellular glutamate pool was by far the least ^{13}C-labeled of those three amino acids: the unlabeled mass isotopomer (M+0) made up 73 % of the amino acid, thereby being reduced 1.3-fold compared to the naturally labeled control (Figure 25C). While also the single labeled mass isotopomer (M+1) was slightly enriched, fully labeled glutamate (M+5) contributed only 9 % to the proteinogenic pool (Figure 25C). Considering the obvious tight regulation of amino acid biosynthesis observed for the majority of amino acids including glycine and serine, exogenous glutamate added to the medium might be the precursor of the intracellular alanine, aspartate, and glutamate pools. On the other hand, labeling studies with acetate and glycerol as well as glucose showed that those three amino acids were derived through *de novo* biosynthesis from the main carbon source rather than amino acids added to the medium (Figure 15, Figure 18). This could indicate that vegetable oil might serve as molecular precursor for glutamate, aspartate, and alanine.

In order to resolve the uncertain molecular origin of glutamate, aspartate, and alanine more closely, [$^{13}C_5$] glutamate was added to the culture medium to replace the naturally labeled amino acid. Close inspection of the cell protein formed under these conditions revealed only a minor contribution of the exogenous glutamate to the intracellular proteinogenic pools (Figure 23, Figure 26, appendix Table 18). The added glutamate contributed merely 12 % to the respective intracellular pool. This was highlighted by the small fraction of fully labeled glutamate from cell protein: only 8 % displayed the M+5 mass isotopomer. Unlabeled glutamate made up 78 % of the intracellular pool (Figure 26C). Aspartate was even less ^{13}C-labeled with a SFL_{corr} of 5 %. Here, the unlabeled mass isotopomer was only 1.1-fold reduced compared to the naturally labeled control. While the single (M+1) and double labeled (M+2) mass isotopomers were elevated, the fully ^{13}C-labeled aspartate was only slightly enriched (Figure 26B). Alanine showed a ^{13}C enrichment of 2 % (appendix Table 18) and exhibited almost the same mass isotopomer distribution as the naturally labeled control (Figure 26A). These data confirmed that, while glutamate is taken up by the cell and present in large amounts in the medium (Figure 13B), de novo biosynthesis of the amino acid seems to be active. In S. cerevisiae, α-ketoglutarate is reduced to glutamate via the NADP-dependent glutamate dehydrogenase, encoded by GDH1 or GDH3, assimilating ammonia (Magasanik, 2003). Interestingly, A. gossypii only contains the NAD-dependent isoform of this enzyme, encoded by the gene GDH2, catalyzing the reverse reaction: glutamate oxidation to α-ketoglutarate (Gomes et al., 2014; Ribeiro et al., 2012). The anabolic formation of glutamate, therefore, relies on transamination reactions, e.g. of α-ketoglutarate with glutamine as nitrogen donor (Guillamón et al., 2001). This reaction is catalyzed by the enzyme glutamate synthase (encoded by the gene GLT1), which was described for S. cerevisiae (Guillamón et al., 2001) and is annotated for A. gossypii (Gomes et al., 2014). In addition to glutamate, also alanine and aspartate are connected to the carbon core metabolism by a single transamination step. The enzyme alanine transaminase (encoded by the gene ALT) catalyzes the transamination of pyruvate to alanine with the nitrogen donor glutamate (Escalera-Fanjul et al., 2017; García-Campusano et al., 2009). While S. cerevisiae contains two isoforms of this enzyme (ALT1 and ALT2) (Escalera-Fanjul et al., 2017), only one alanine transaminase is annotated for A. gossypii (Förster et al., 2014). The aspartate transaminase (encoded by the gene AAT) converts the metabolic precursor oxaloacetate into aspartate while glutamate serves as nitrogen donor (Morin et al., 1992). Two isoforms, a cytosolic and a mitochondrial enzyme, are present in S. cerevisiae (Morin et al., 1992; Yagi et al., 1982) and were also annotated for A. gossypii (Förster et al., 2014). The reason for the deregulated de novo biosynthesis of glutamate and to a lesser extent also aspartate and alanine might lie in their central role as reactant in cellular transamination reactions. The key player glutamate accounts for 85 % of the cellular nitrogen demand (Ljungdahl and Daignan-Fornier, 2012), contributing its amino-group mostly to other amino acids. Glutamine, on the other hand, donates about 15 % of its amide nitrogen to amino acids as well as purines and pyrimidines (Ljungdahl and Daignan-Fornier, 2012). Alanine and also aspartate are derived from pyruvate or oxaloacetate, respectively, by a one-step transamination reaction (Escalera-Fanjul et al., 2017; Yagi et al., 1982). Considering the importance of nitrogen in the cellular metabolism, it seems obvious that those reactions are not strongly feedback regulated. The fact that alanine was by far the most labeled of the three, when [U^{13}C] yeast extract was fed, could be attributed to the absence of the enzyme alanine-glyoxylate aminotransferase, encoded by *AGX1* (Kato and Park, 2006). This reduces the number of possible transaminating reactions, alanine could be involved in.

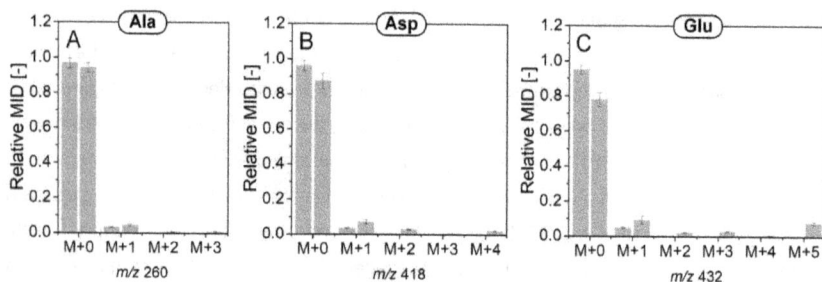

Figure 26: Relative mass isotopomer distributions (MIDs) of proteinogenic alanine (A), aspartate (B), and glutamate (C) from *A. gossypii* B2 grown on complex medium with rapeseed oil as main carbon source. MIDs are presented for medium with naturally labeled substrates (grey) and [^{13}C$_5$] glutamate (light blue). Data were obtained from three individual replicates.

In order to gain a deeper understanding of the metabolism of *A. gossypii* on rapeseed oil, the ^{13}C-labeling incorporation of glycogen-derived cellular glucose was also inspected. Fully ^{13}C-labeled yeast extract as well as [^{13}C$_5$] glutamate resulted both in a SFL$_{corr}$ of 6 % for glucose, respectively (Figure 23, Table 18). The single labeled mass isotopomer (M+1) was 3-fold increased compared to the naturally labeled compound (Table 18). The majority of cellular glucose, however, originated from vegetable oil (Figure 23).

These data showed that glutamate plays a central role in the metabolism of *A. gossypii*. The fact that it is taken up in large amounts during the growth phase (6 g L^{-1}), but contributes only slight ^{13}C enrichment to proteinogenic amino acids, obviously suggests strong conversion through the TCA cycle. Evidently, glutamate plays a central role in nitrogen metabolism, which would be very interesting to investigate further, especially with regard to riboflavin that contains a total of four nitrogen atoms. The addition of fully ^{13}C-labeled yeast extract to the medium revealed that most amino acids from the yeast extract were efficiently taken up and incorporated into cell protein. Thus, with the exception of alanine, aspartate, and glutamate, *A. gossypii* exhibited a tight regulation of *de novo* biosynthesis of amino acids.

4.4.5   Carbon fluxes of *A. gossypii* B2 during growth on vegetable oil

The previous chapters gave a detailed overview of the origin of growth associated building blocks. The supplemented yeast extract as well as glycine contributed greatly to the formation of proteinogenic amino acids, while the impact of extracellular glutamate was reduced in comparison (Figure 23). These ^{13}C tracer studies also allowed the conclusion that rapeseed oil, the main carbon source of the process, participated in the biosynthesis of a few selected amino acids, like glutamate, and the formation of cellular glycogen. Taken together this indicated, at least on an additive level, that the TCA cycle was active (glutamate conversion) as also were the glyoxylate shunt and the gluconeogenic route (oil conversion into glycogen). On the basis of these ^{13}C labeling data and interpreting the individual labeling information plus measured extracellular fluxes from the different parallel ^{13}C tracer studies, carbon fluxes were calculated with respect to rapeseed oil as major carbon source. For that, the demand of anabolic precursors for cell growth had to be derived and converted into rates. The theoretical precursor demand for *A. gossypii* was accessible from literature data: the genome-scale metabolic model of *A. gossypii* provided most of the information for the biomass composition of the fungus (Ledesma-Amaro et al., 2014a). The cellular demand for acetyl-CoA and glyceraldehyde 3-phosphate, the building blocks of lipids, was adjusted to growth on vegetable oil. While growth on glucose leads to a lipid content of about 0.08 to 0.15 g g_{CDW}^{-1}, growth on vegetable oil results in a lipid content of up to 0.22 g g_{CDW}^{-1} (Stahmann et al., 1994). The anabolic precursor demand for the riboflavin production during growth was derived from the literature and KEGG via the underlying pathway stoichiometry (Bacher et al., 2000; Kanehisa et al., 2017; Kanehisa and Goto, 2000; Kanehisa et al., 2016). The final values for the eleven anabolic precursors are listed in Table 12. The demand assumes that all cellular building blocks are synthesized *de novo*. However, the results in the previous chapter demonstrated that a large fraction of cellular protein and glycogen was covered by mere uptake of amino acids and yeast extract supplemented in the medium (Figure 23). This was used to correct the demand into the real values remaining under the given condition. The uptake of each amino acid was calculated by addition of the respective SFL_{corr} for that amino acid from the single isotope experiments. The remaining fraction of the amino acid was then attributed to *de novo* synthesis from vegetable oil. Each cellular building block, e.g. amino acids or nucleotides, is derived from one or more out of a total of eleven precursors (appendix Table 19). The demand for a single precursor is therefore the sum of the cellular demand of its building blocks.

Pyruvate, e.g., is the metabolic precursor of alanine, valine, isoleucine, and leucine (appendix Table 19). However, since most of these amino acids were either fully or partially taken up by the cell from the culture medium, the *de novo* pyruvate demand (1783.1 µmol g_{CDW}^{-1}) was tremendously decreased to 71.5 µmol g_{CDW}^{-1} when ^{13}C labeling data were taken into account. Thus, the experimental *de novo* precursor demand for growth on complex medium differed greatly from the values (Table 12), underlining the importance of taking ^{13}C labeling data into consideration as well as the impact of the medium to support intracellular pools. Definition of metabolite balances and flux calculations leading to the final carbon flux distribution are given in the appendix (Chapter 6).

Table 12: Estimation of the de novo demand for anabolic precursors during growth of *A. gossypii* on complex medium and rapeseed oil. First, the total precursor demand (total demand) was taken from literature (Ledesma-Amaro et al., 2014a) and adjusted for growth on vegetable oil based on (Stahmann et al., 1994). The total demand for each anabolic precursor was covered by two routes: (i) uptake of external building blocks from complex ingredients, which biosynthetically originate from the respective precursor (e.g. alanine, valine, etc. from pyruvate) and (ii) *de novo* synthesis of the building blocks from vegetable oil. Correlation of the total demand values with experimental summed fractional labeling (SFL_{corr}) data from combined results of parallel ^{13}C isotope studies with [$^{13}C_2$] glycine, [^{13}C] formate, [$^{13}C_5$] glutamate, and [$U^{13}C$] yeast extract (appendix Table 17 and Table 18) yielded the measured percentage of (i) the anabolic precursor that could be neglected due to the uptake of advanced metabolites (e.g. amino acids, nucleotides) and (ii) the resulting percentage of de novo precursor demand for growth on complex medium and rapeseed oil, which was then converted into the resulting de novo demand. The full length bar indicates the individual contributions visually: the purple fraction depicts the percentage covered from complex ingredients, while the grey fraction depicts the resulting de novo biosynthetic fraction of the precursor. The *de novo* demand for acetyl CoA was not specified, which is explained in more detail in Table 20. 3PG, 3-phosphoglycerate; AcCoA, acetyl-CoA; AKG, α-ketoglutarate; E4P, erythrose 4-phosphate; F6P, fructose 6-phosphate; G3P, glyceraldehyde 3-phosphate; G6P, glucose 6-phosphate; OAA, oxaloacetate; PEP, phosphoenolpyruvate; PYR, pyruvate; R5P, ribose 5-phosphate.

Anabolic precursor	Total demand		Uptake[a]	De novo biosynthesis	Resulting de novo demand
	[µmol g_{CDW}^{-1}]		[%]	[%]	[µmol g_{CDW}^{-1}]
G6P	604.8		12.5 ± 0.4	87.5 ± 0.4	529.2 ± 2.7
F6P[b]	821.0		0.0	100.0	821.0
R5P[b]	329.9		94.8	5.2	17.3
E4P	238.8		98.0 ± 2.9	2.0 ± 3.4	4.8 ± 7.2
G3P[b]	240.4		0.0	100.0	240.4
3PG	707.4		94.8 ± 5.1	5.2 ± 5.0	37.1 ± 36.0
PEP	449.6		97.9 ± 2.9	2.1 ± 2.7	9.6 ± 14.3
PYR	1783.1		96.0 ± 1.0	4.0 ± 1.0	71.5 ± 14.9
AcCoA	6572.7		n.d.	n.d.	n.d.
OAA	1124.5		88.5 ± 1.9	11.5 ± 1.9	129.8 ± 20.5
AKG	800.4		60.8 ± 1.5	39.2 ± 1.5	313.7 ± 10.7

[a] Percentage negligible due to uptake of advanced building blocks, i.e. amino acids or nucleotides, from the medium.
[b] The demand for the precursor is assumed. Therefore, no standard deviation could be calculated for the according values

Growth of *A. gossypii* on vegetable oil and complex medium resulted in a strong carbon flux through the ß-oxidation pathway as well as TCA cycle (Figure 27). Vegetable oil was cleaved into three fatty acids (average chain length of 17.3 carbon atoms, denoted as FA in Figure 27) (Stahmann et al., 1994) and glycerol in the extracellular space (Stahmann et al., 1997). Fatty acids were then taken up by the cell (0.43 mmol $g_{CDW}^{-1}h^{-1}$) and oxidized to acetyl-CoA through the ß-oxidation pathway, located in the peroxisome (Shen and Burger, 2009; Vorapreeda et al., 2012). The carbon flux from acetyl-CoA into biomass (0.506 mmol g^{-1} h^{-1}) assumes that the intracellular storage lipids have to be synthesized *de novo*, which is, however, unlikely as described below. A small fraction of acetyl-CoA, i.e. 0.141 mmol g^{-1} h^{-1}, was then assimilated through the glyoxylate shunt, yielding an overall carbon flux into malate of 0.141 mmol g^{-1} h^{-1}. The majority of acetyl-CoA, however, was channeled into the TCA cycle, where it was decarboxylated and finally contributed to cellular respiration. The highly reduced fatty acids are oxidized in order to become accessible for the cell. The strong decarboxylation of acetyl-CoA in the TCA cycle is balanced by oxygen consumption during cellular respiration. These facts match the typically observed RQ of < 1 for growth on lipids (Stahmann et al., 1994), which is characterized by high oxygen consumption and low carbon dioxide formation. Nevertheless, the high flux through the TCA cycle seemed surprising at first, given the low biomass yield of 186 g mol^{-1} and growth rate (0.08 h^{-1}) (Table 6) and the resulting low demand of cellular ATP. However, the high TCA cycle activity might indicate a high maintenance demand of the cell. In a ^{13}C metabolic flux study with *Sorangium cellulosum*, a slow growing myxobacterium, a high carbon flux through the TCA cycle could also be observed, which was potentially linked to a high maintenance metabolism (Bolten et al., 2009). Wild type *B. subtilis* also requires a lot of energy for its maintenance (Tännler et al., 2008a). However, knockout of the terminal cytochrome *bd* oxidase in the respiratory chain of a riboflavin producing *Bacillus* strain, resulted in more efficient proton pumping, thereby decreasing the high maintenance metabolism, and thus, even improving riboflavin production performance (Zamboni et al., 2003). Therefore, the targeted reduction of the maintenance demand might also display a potential metabolic target and improve production performance in *A. gossypii*.

The acetyl-CoA that was channeled through the TCA cycle plus the oxaloacetate from the glyoxylate shunt together formed citrate, which was then decarboxylated into α-ketoglutarate (Figure 27). The large flux from α-ketoglutarate to intracellular glutamate (0.389 mmol g^{-1} h^{-1}) ensured the observed ^{13}C enrichment of glutamate (combined SFL_{corr} of 26.5 %), which was largely naturally labeled, given the strong uptake of extracellular glutamate during growth (0.142 mmol g^{-1} h^{-1}). The intracellular glutamate pool fed into biomass: while Figure 27 only shows lumped reactions, Figure 28A displays a more detailed picture.

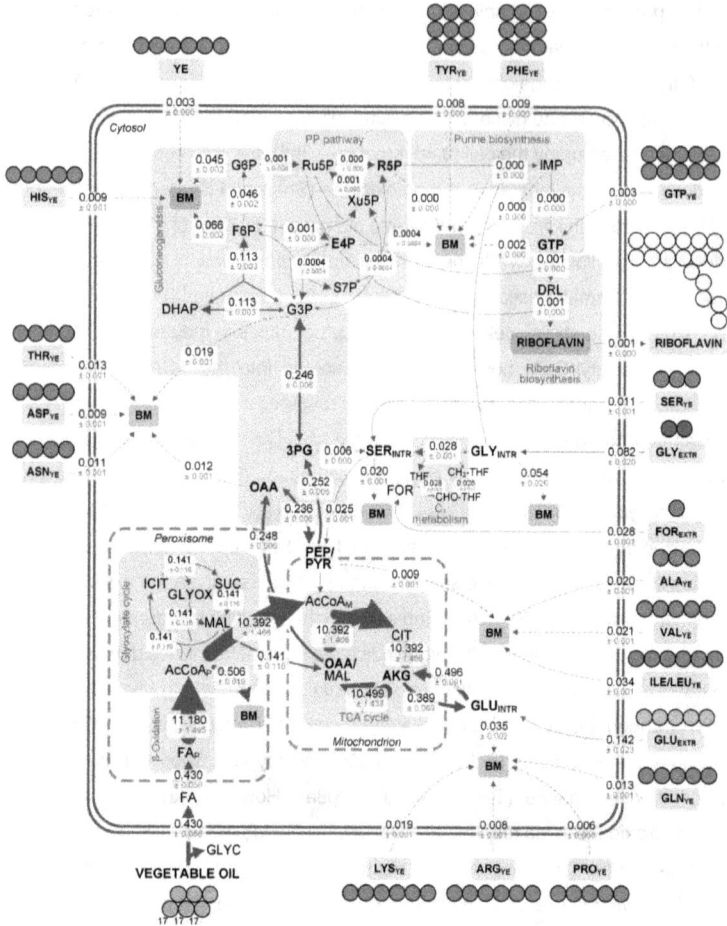

Figure 27: Carbon fluxes during growth of *A. gossypii* B2 on rapeseed oil and complex medium. Fluxes are given in mmol g_{CDW}^{-1} h^{-1} and are normalized to the substrate uptake rate (0.430 mmol g^{-1} h^{-1}). Flux calculations were derived from parallel ^{13}C tracer studies with [$^{13}C_2$] glycine, [^{13}C] formate, [$^{13}C_5$] glutamate, and [$U^{13}C$] yeast extract (appendix Table 17, Table 18). The arrow thickness is proportional to the corresponding flux. The direction of net fluxes is indicated by size of arrow head. Dashed arrows represent fluxes into biomass. Reactions at the OAA/MAL and PEP/PYR node could not be resolved by this approach and represent lumped fluxes. The rate of the pyruvate dehydrogenase could not be determined. In order to simplify the figure, only GTP uptake is depicted as immediate riboflavin precursor. Other nucleotides as well as cysteine and methionine, which are also taken up from the medium, are not shown. Note that the conversion of citrate to isocitrate via aconitase most likely does not occur in the peroxisome (Murakami and Yoshino, 1997). 3PG, 3-phosphoglycerate; AcCoA$_{P/M}$, peroxisomal/mitochondrial acetyl-CoA; AKG, α-ketoglutarate; ALA, alanine; ARG, arginine; ASN, asparagine; ASP, aspartate; BM, biomass; CH$_2$-THF, 5,10-methylenetetrahydrofolate; CHO-THF, 10-formyltetrahydrofolate; CIT, citrate; DHAP, dihydroxyacetone phosphate; DRL, 6,7-dimethyl-8-ribityllumazine; E4P, erythrose 4-phosphate; F6P, fructose 6-phosphate; FA, fatty acids (here three FA with an average chain length of 17.3 carbon atoms); FOR, formate; G3P, glyceraldehyde 3-phosphate; G6P, glucose 6-phosphate; GAR, glycineamide ribonucleotide; GLN, glutamine; GLU, glutamate; GLY, glycine; GLYC, glycerol; GLYOX, glyoxylate; GTP, guanosine triphosphate; IMP, inosine monophosphate; MAL, malate; OAA, oxaloacetate; PEP, phosphoenolpyruvate; PYR, pyruvate; R5P, ribose 5-phosphate; Ru5P, ribulose 5-phosphate; S7P, sedoheptulose 7-phosphate; SER, serine; THF, tetrahydrofolate; Xu5P, xylulose 5-phosphate; YE, yeast extract.

76

Carbon fluxes into the single proteinogenic amino acids illustrate the strong regulation of their biosynthesis: intracellular glutamate only supplied small fractions of the respective amino acids, while most of their demand was covered through uptake from the medium (Figure 28A). The strong backflux from glutamate to α-ketoglutarate resulted in an α-ketoglutarate-synthesizing netflux (0.107 mmol g^{-1} h^{-1}). Reactions involving the OAA/malate pool as well as the PEP/pyruvate pool could not be resolved by this approach and were calculated as lumped reactions, i.e. a net flux from oxaloacetate to the combined PEP/pyruvate pool (0.236 mmol g^{-1} h^{-1}). Oxaloacetate is an important anabolic precursor, contributing to the synthesis of various amino acids as well as pyrimidine biosynthesis (appendix Table 19). The detailed distribution of carbon fluxes into biomass starting from the precursor oxaloacetate is presented in Figure 28B. Intracellular aspartate is derived from extracellular aspartate and to similar extent from oxaloacetate. However, there was only a small flux of 0.003 mmol g^{-1} h^{-1} from intracellular aspartate to asparagine and threonine, on the other hand, indicating strong uptake of these two amino acids from the medium (fluxes of 0.011 mmol g^{-1} h^{-1} and 0.013 mmol g^{-1} h^{-1}, respectively) (Figure 28B). The carbon fluxes through the upper metabolism, i.e. gluconeogenesis and PP pathway were comparatively low. This can be attributed to the high uptake of amino acids from the culture medium and the resulting low *de novo* precursor demand. Here, the precursor demand for ribulose 5-phosphate, ribose 5-phosphate, and erythrose 4-phosphate was almost completely covered by the reactions of the non-oxidative part of the PP pathway (Figure 27). A small carbon flux of 0.001 mmol g^{-1} h^{-1} was assigned to the oxidative branch. This raises the question of NADPH supply in the cell, since the enzymes glucose-6-phosphate dehydrogenase as well as phosphogluconate dehydrogenase are often regarded as main donors for cytosolic NADPH (Panagiotou et al., 2009). Even if assuming that all extra carbon from the non-oxidative part of the PP pathway was channeled through gluconeogenesis and subsequently into the oxidative part of the PP pathway, the resulting flux would have been only 0.002 mmol g^{-1} h^{-1}. Lipid biosynthesis requires large amounts of NADPH, however, storage lipids in *A. gossypii* are most likely not synthesized *de novo*, but formed through transesterification with fatty acids and glycerol taken up from the culture supernatant. It was reported that the lipid composition of *A. gossypii* grown on vegetable oil depended on and resembled the composition of the vegetable oil used as a substrate (Stahmann et al., 1994). Another NADPH-requiring enzyme during growth, glutamate dehydrogenase (*GDH1* or *GDH3*), is absent in *A. gossypii* (Gomes et al., 2014; Ribeiro et al., 2012) and thus, glutamate *de novo* synthesis occurs via the enzyme glutamate synthase (Chapter 4.4.4). Two isoforms of this enzyme exist, a NAD-dependent and a NADP-dependent form (EC 1.4.1.14 and EC 1.4.1.13, respectively) and are both annotated for *A. gossypii* as well as *S. cerevisiae* in the KEGG database (Kanehisa et al., 2017; Kanehisa and Goto, 2000; Kanehisa et al., 2016). While there is no experimental evidence on cofactor specificity for

A. gossypii, for *S. cerevisiae*, however, the enzyme was reported to be NAD-dependent (Guillamón et al., 2001; Vanoni and Curti, 2008). The glutamate synthase of the yeast *Kluyveromyces marxianus*, on the other hand, was proven to use NADP as cofactor (Nisbet and Slaughter, 1980). Thus, further studies are needed to unravel the cofactor requirement for glutamate biosynthesis in *A. gossypii*. Given the above mentioned data and the fact that most amino acids were readily taken up from the medium, the NADPH requirement of the cell is comparably low. Nonetheless, there are different potentially NADPH-generating enzymes in *A. gossypii*: isocitrate dehydrogenase (IDH), malic enzyme (MaE), and NADH kinase. In *A. gossypii* there are two isoforms of the IDH, which converts isocitrate into α-ketoglutarate: the mitochondrial isoenzyme is NAD-specific, whereas the NADP-dependent enzyme is localized to the peroxisome (Maeting et al., 2000), where it is thought to be involved in cofactor regeneration during oxidation of unsaturated fatty acids via 2,4-dienoyl-CoA reductase (Maeting et al., 2000). Three isoenzymes were detected in *S. cerevisiae*, two of which are localized to the mitochondria (NAD- and NADP-specific) and a third one, NAD-specific, is located in the cytosol (Haselbeck and McAlister-Henn, 1993). The NADP-dependent decarboxylation of malate to pyruvate, catalyzed by MaE, was confirmed for *Aspergillus niger* (Jernejc and Legisa, 2002). The MaE from *S. cerevisiae* has been reported to use both NAD as well as NADP as cofactors (Boles et al., 1998). The enzyme is annotated for *A. gossypii* (KEGG, AGL068W), however, no experimental data for the cofactor specificity are available in the literature (Kanehisa et al., 2017; Kanehisa and Goto, 2000; Kanehisa et al., 2016). Another interesting NADPH-providing reaction is the phosphorylation of NADH by NADH kinase. This reaction is located in the mitochondria and was described, among others, for *S. cerevisiae* and *Aspergillus nidulans* (Outten and Culotta, 2003; Panagiotou et al., 2009) and is annotated for *A. gossypii* (KEGG, AFL063W) (Kanehisa and Goto, 2000).

The flux distribution presented here shows that the glyoxylate shunt plays an important role in carbon assimilation during growth on vegetable oil, which is then channeled into gluconeogenesis. The decreased carbon flux through the upper metabolism, i.e. gluconeogenesis and PP pathway, reflects the strong regulation of amino acid *de novo* biosynthesis due to the ample supply of amino acids as well as nucleotides in the culture supernatant (Zhang et al., 2003). While pioneering ^{13}C flux studies used ethanol or glucose as carbon sources on semi-defined media (de Graaf et al., 2000; Jeong et al., 2015), this is the first report of a carbon flux distribution with *A. gossypii* on vegetable oil and complex medium during growth. The concerted action of the main carbon source vegetable oil and the complex substrate yeast extract regarding the formation of cell building blocks, illustrated by the flux map (Figure 27), highlights the complexity of the given process.

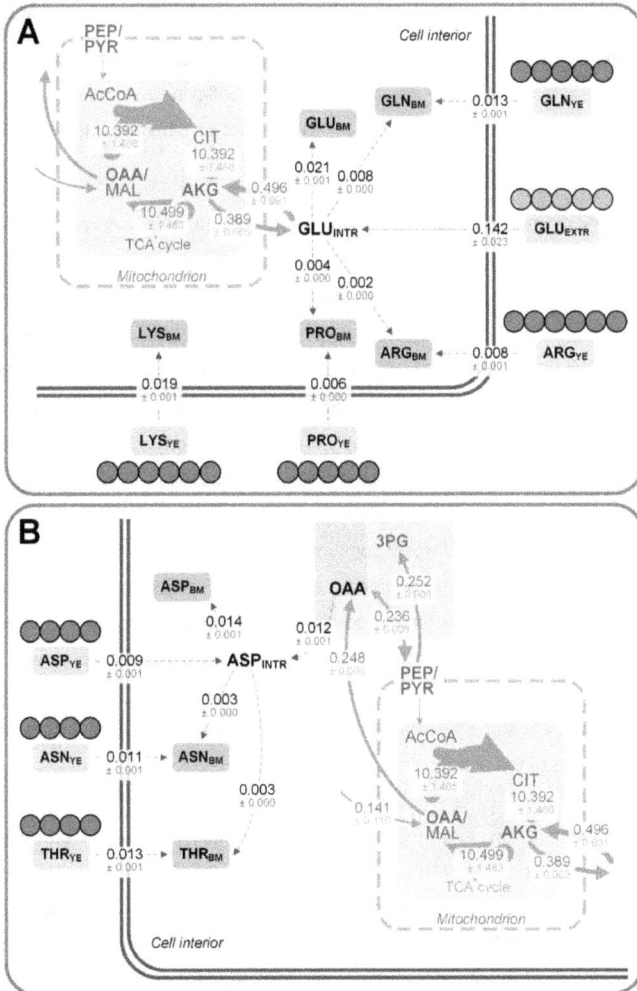

Figure 28: Detailed view of carbon fluxes into biomass for amino acids stemming from glutamate (A) or aspartate (B) during growth of *A. gossypii* B2 on vegetable oil and complex medium. Fluxes are given in mmol g$_{CDW}^{-1}$ h^{-1} and are normalized to the substrate uptake rate (0.430 mmol g^{-1} h^{-1}). Flux calculations were derived from parallel ^{13}C tracer studies with [^{13}C$_2$] glycine, [^{13}C] formate, [^{13}C$_5$] glutamate, and [U^{13}C] yeast extract (appendix Table 17, Table 18). The arrow thickness is proportional to the corresponding flux. The direction of net fluxes is indicated by size of arrow head. Dashed arrows represent fluxes into biomass. Reactions at the OAA/MAL and PEP/PYR node could not be resolved by this approach and represent lumped fluxes. The rate of the pyruvate dehydrogenase could not be determined. The contribution of intracellular glutamate or aspartate to glutamine and asparagine, respectively, could not be measured and was estimated based on the ^{13}C labeling data for proline and threonine. The full figure is presented in Figure 27. 3PG, 3-phosphoplycerate; AcCoA, acetyl-CoA; AKG, α-ketoglutarate; ALA, alanine; ARG, arginine; ASN, asparagine; ASP, aspartate; BM, biomass; CIT, citrate; GLN, glutamine; GLU, glutamate; MAL, malate; OAA, oxaloacetate; PEP, phosphoenolpyruvate; PRO, proline; PYR, pyruvate; THR, threonine; YE, yeast extract.

4.5 Unraveling the building blocks of riboflavin using LC/MS and ^{13}C NMR

After successful determination of carbon fluxes in the growth phase, the riboflavin production phase was investigated next. Since labeling analysis of riboflavin was not feasible with GC/MS, LC/MS and NMR were employed to quantify the impact of different tracers on the carbon skeleton of the vitamin.

4.5.1 LC/MS measurements confirm glycine and formate as riboflavin building blocks

An initial experiment verified that the mass isotopomer distribution of accumulated and partially purified riboflavin from a culture of *A. gossypii* B2 was accessible by LC/MS (Figure 29). Using naturally labeled medium ingredients, samples from the end of the production process after 144 h were harvested, partially purified (Chapter 3.4.3) and analyzed via LC/MS. The observed mass isotopomer distribution of the vitamin, corrected for natural isotope abundance, exhibited the expected monoisotopic mass. Accordingly, the incorporation of individual tracers into the product could be quantified by this approach. Further adaptions in the labeling strategy were conducted in order to ensure that each tested precursor was still available during the production phase. Formate was largely consumed prior to riboflavin biosynthesis. Therefore, [^{13}C] formate was added (i) at the beginning and (ii) in a parallel experiment after 48 h of cultivation. The latter experiment should compensate for eventual formate loss and dilution effects during the preceding growth phase, in order to ensure that the tracer was present at the delayed start of riboflavin production (Figure 13A). The tracer [^{13}C$_2$] glycine, which was present in the medium in high concentrations at all times (Figure 13B), had to be supplemented only at the beginning of the experiment.

Riboflavin synthesized on fully ^{13}C-labeled glycine was M+2 enriched, whereas the monoisotopic mass isotopomer (M+0) was strongly decreased (Figure 29A). This indicated the incorporation of glycine as two-carbon unit mainly. A small fraction was attributed to the M+1 isotopomer. In contrast to M+0 and the M+2 mass isotopomer, the small M+1 peak suffered from a rather low signal quality of the underlying peak in the mass spectrometric analysis. It was therefore not fully clear, to which extent it represented the incorporation of a single ^{13}C atom from the tracer or simply background interference. From the observed MID a summed fractional labeling of 9 % could be calculated for the total molecule. Considering that the maximum SFL for M+2 in riboflavin from [^{13}C$_2$] glycine equals 12 %, this corresponded to a total fraction of 79 % riboflavin molecules, which contained an entire glycine residue. The contribution of formate on riboflavin synthesis was strongly dependent on the point in time of tracer addition. When formate was added after 0 h of cultivation, formed riboflavin was so slightly enriched in ^{13}C that it disappeared in signal background (Table 13, appendix Table 22 and Table 23). This suggested that formate, initially present in the medium, had been taken up and metabolized differently prior to riboflavin biosynthesis. Measurement of the formate concentration in the culture supernatant confirmed this assumption (data not shown). Thus, extracellular formate was not available during main riboflavin production. The ^{13}C integration of formate was 100-fold greater, when the tracer was added after 48 h (Table 13). Here, the M+1 fragment in riboflavin was enriched, indicating that formate contributed to only one carbon atom of the vitamin (Figure 29B).

Table 13: Summed fractional labeling (SFL) of riboflavin, derived from LC/MS based measurement of the product, extracted after 144 h from culture broth of *A. gossypii* B2, grown on different ^{13}C tracers and naturally labeled complex medium with rapeseed oil. The SFL, which was corrected for naturally occurring isotopes, was determined using the original measurement data (appendix Table 22, Table 23), and was corrected for the dilution of labeling through naturally labeled pre-culture medium at the beginning of the cultivation (SFL$_{corr}$). The right column displays the values for the respective contribution of the used ^{13}C-labeled precursor, which also takes into consideration the possible carbon transition of the molecules. The correction is described in Chapter 3.5.2. Data denote mean values of three independent replicates.

^{13}C Tracer	Time of addition	SFL	SFL$_{corr}$	Contribution Ψ
	[h]	[%]	[%]	[%]
Nat. lab. precursor	0	0.0 ± 0.0	0.0 ± 0.0	0.0 ± 0.0
[$^{13}C_2$] Glycine	0	9.0 ± 0.8	9.3 ± 0.8	79.1 ± 6.8
[^{13}C] Formate	0	0.0 ± 0.0	0.0 ± 0.0	0.0 ± 0.0
	48	1.0 ± 0.1	1.0 ± 0.1	17.0 ± 1.7
[3-^{13}C] Serine	0	0.2 ± 0.0	0.3 ± 0.0	5.1 ± 0.2
	48	0.4 ± 0.0	0.4 ± 0.0	6.8 ± 0.3
[$^{13}C_3$] Serine	0	1.2 ± 0.1	1.6 ± 0.1	9.1 ± 0.6
	48	1.9 ± 0.2	2.0 ± 0.2	11.3 ± 1.1
[U^{13}C] Yeast extract	0-32	4.5 ± 0.6	5.1 ± 0.7	n.d.
	32	8.0 ± 1.1	9.0 ± 1.2	n.d.

4.5.2 NMR analyses yield higher resolution of glycine and formate as precursors

Additional ^{13}C experiments were run in order to resolve the molecular origin of the carbon atoms in riboflavin even more precisely using NMR analysis. This offered the great advantage that not only the number of labeled carbon atoms in riboflavin became accessible, but also the positional ^{13}C enrichment. The latter appeared particularly valuable to clarify the single carbon incorporations, which could not be resolved by LC/MS. Prior to ^{13}C labeling experiments, purity and structure of riboflavin were confirmed by two-dimensional NMR using commercially available riboflavin and naturally labeled riboflavin, recovered after 144 h from culture broth and subsequently purified. A ^1H NMR spectrum in the first and a ^{13}C NMR spectrum in the second dimension provided identical results (data not shown). Every peak could be attributed to exactly one carbon atom. The peak areas for each carbon were identical, reflecting natural ^{13}C abundance. A 1D ^{13}C NMR spectrum for naturally labeled riboflavin from *A. gossypii* is presented in Figure 29D.

For cultivations preceding NMR analysis, the same tracer set-up was chosen as for LC/MS measurements: fully ^{13}C-labeled glycine added after 0 h, and ^{13}C-labeled formate added after (i) 0 h and (ii) 48 h in parallel experiments. The ^{13}C NMR data of riboflavin, which were corrected for natural ^{13}C labeling, are shown in Table 14 (raw data: appendix Table 24). Initial ^{13}C tracer enrichment was monitored and relative enrichments were corrected accordingly (Chapter 3.5.2). Riboflavin synthesized on medium, containing labeled glycine, yielded a 72 % positional enrichment at the carbons C_{4a} (136.8 ppm chemical shift) and C_{10a} (150.8 ppm chemical shift) in the pyrimidine ring (Table 14). Both peaks appeared as duplets due to ^{13}C coupling of the neighboring carbon atoms (Figure 29G). This demonstrated that glycine was incorporated as entire two-carbon unit, which confirmed previous LC/MS results (Figure 29A, Table 13). There was no ^{13}C enrichment at 155.5 ppm representing carbon atom C_2, known to originate from folate-dependent one-carbon metabolism. This confirmed that the M+1 peak, which was observed in the LC/MS spectrum of riboflavin, could be attributed to background noise (Figure 29A). Addition of [^{13}C] formate after 0 h and 48 h of cultivation yielded 4 % and 12 % ^{13}C enrichment at the C_2 atom, respectively (Figure 29F, Table 14). Obviously, formate provided a single carbon atom to the vitamin.

Several decades ago, glycine was identified as the donor of the neighboring carbons C_{4a} and C_{10a} in the pyrimidine ring of riboflavin (Plaut, 1954a). This pioneering study also tried to elucidate the contribution of glycine to the carbon C_2 in the ring, which would have indicated the presence or absence of a glycine cleavage system, existing in yeasts such as *S. cerevisiae* (Piper et al., 2002). At the time, this could, however, not be proven due to experimental noise at the resulting low labeling enrichment. The genome annotation shows *GCV1* as a missing gene for a glycine cleavage system (Gomes et al., 2014). However, until now there was no experimental proof for the absence or presence of a glycine cleavage system in *A. gossypii* apart from the gene annotations. Here, fully ^{13}C-labeled glycine was incorporated into riboflavin exclusively at two carbon positions: the neighboring carbons C_{4a} and C_{10a} in the pyrimidine ring (Figure 29G, Figure 30A). In contrast, the carbon C_2 did not receive any ^{13}C label, while the labeled tracer was present throughout. The enrichment was exactly at the abundance of riboflavin from a non-labeled control experiment (Figure 29G) and also from a commercial standard (data not shown). It can be concluded from the NMR data that *A. gossypii* does not operate a glycine cleavage system. Thus, glycine cannot act as C_1 donor. However, glycine contributes to the majority of the two-carbon unit in riboflavin. In the tracer study with fully labeled glycine, the two atoms C_{4a} and C_{10a} exhibited a positional enrichment of 72 % (Figure 30A).

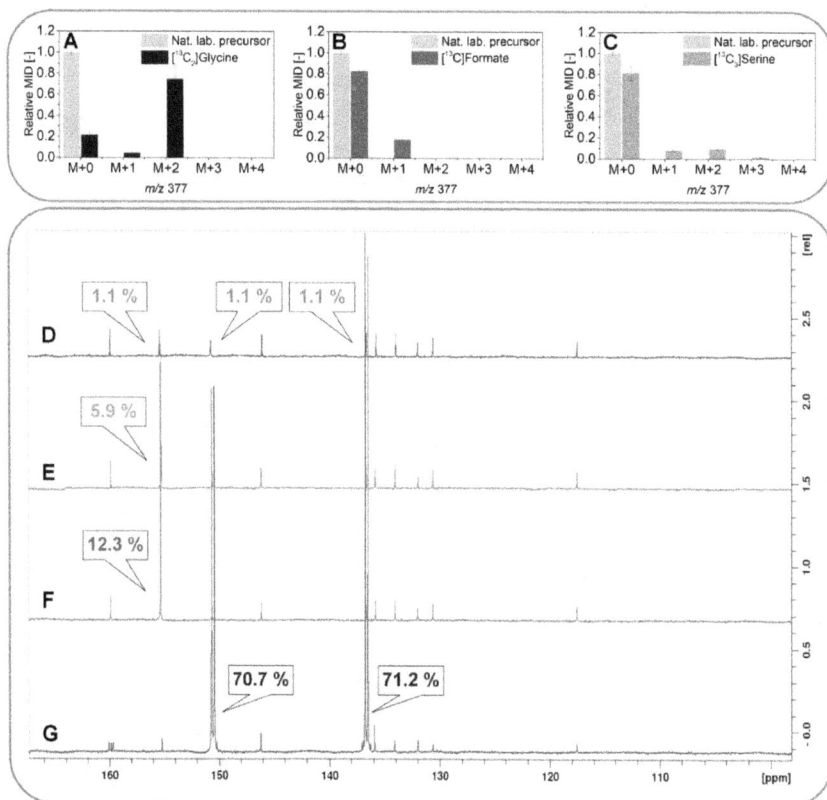

Figure 29: LC/MS spectra of riboflavin (*m/z* 377) (A-C) and proton-decoupled ^{13}C NMR spectra of riboflavin (D-G) derived from cultures of riboflavin producing *A. gossypii* B2 grown on $[^{13}C_2]$ glycine (A, G), $[^{13}C]$ formate (B, F), $[^{13}C_3]$ serine (C), $[3-^{13}C]$ serine (E), and on naturally labeled precursors (D). Samples were harvested after 144 h. Riboflavin produced on naturally labeled medium is depicted in grey for LC/MS data, which were corrected for naturally occurring isotopes. Labeled glycine was added at the beginning of the cultivation, labeled formate and serine were added after 48 h. As higher mass isotopomers were below the detection limit, only the first five are depicted (A-C). Data were obtained from three independent experiments with a mean standard deviation of 5 % for ^{13}C NMR data if riboflavin.

Table 14: Relative ^{13}C enrichment of all seventeen carbon atoms of riboflavin produced by *A. gossypii* B2 from different ^{13}C-labeled tracer substrates on complex medium with rapeseed oil. Riboflavin was recovered from the culture broth after 144 h. The labeling was analyzed by ^{13}C NMR. The measurement data (uncorrected, appendix Table 24) were corrected values for naturally occurring isotopes and dilution through naturally labeled pre-culture medium. The time refers to the respective tracer addition. Data denote mean values of three independent replicates with a mean standard deviation of 8 %. For, formate; Glu, glutamate; Gly, Glycine; Ser, serine; YE, yeast extract.

C-atom	Chemical shift [ppm]	Nat. lab. precursors	Corrected relative enrichment [%]						
			[13C] For		[13C$_2$] Gly	[3-13C] Ser	[U13C] YE	[13C$_5$] Glu
		0 h	0 h	48 h	0 h	48 h	32 h	0 h
2	155.5	0.0	3.7	11.6	0.0	5.1	16.2	1.1
4	159.9	0.0	0.0	0.0	0.0	0.0	12.1	1.8
4a	136.8	0.0	0.0	0.0	72.1	0.0	19.6	0.0
5a	134	0.0	0.0	0.0	0.0	0.0	4.2	1.9
6	130	0.0	0.0	0.0	0.0	0.0	4.5	2.3
7	137.1	0.0	0.0	0.0	0.0	0.0	3.0	1.8
7α	18.8	0.0	0.0	0.0	0.0	0.0	3.0	1.7
8	146	0.0	0.0	0.0	0.0	0.0	4.0	1.9
8α	20.8	0.0	0.0	0.0	0.0	0.0	4.0	2.6
9	117.4	0.0	0.0	0.0	0.0	0.0	3.1	0.7
9a	132.1	0.0	0.0	0.0	0.0	0.0	3.0	0.3
10a	150.8	0.0	0.0	0.0	71.6	0.0	19.7	0.0
1'	47.3	0.0	0.0	0.0	0.0	0.0	9.9	2.9
2'	68.8	0.0	0.0	0.0	0.0	0.0	8.7	4.1
3'	73.6	0.0	0.0	0.0	0.0	0.0	7.7	3.0
4'	72.8	0.0	0.0	0.0	0.0	0.0	8.1	3.9
5'	63.4	0.0	0.0	0.0	0.0	0.0	7.3	1.6

4.5.3 The dual role of serine in riboflavin biosynthesis

For *A. gossypii*, serine is neither typically added to the cultivation medium, nor is it an immediate precursor of riboflavin. However, since the addition of ^{13}C serine to the growth phase of the fungus gave rise to ^{13}C enrichment in glycine, it was considered to also participate in riboflavin production on some level. Tracer studies similar to the ones with glycine and formate, combining LC/MS, GC/MS, and NMR analyses were conducted in order to elucidate the impact of serine. In parallel studies, [13C$_3$] serine and [3-13C] serine were used as tracers, the latter specifically fueling the one-carbon metabolism with the ^{13}C label. The LC/MS spectrum of riboflavin, produced on fully labeled serine, showed a decreased monoisotopic mass isotopomer (Figure 29C). At the same time, M+1 and M+2 isotopomers were enriched as well as, however to a lesser extent, M+3. Consequently, serine probably functioned as glycine donor (M+2), C$_1$ donor (M+1), and a combination of the two. GC/MS based analysis of the culture supernatant during the riboflavin biosynthetic phase revealed that exogenous serine contributed to about 8 % of the extracellular glycine pool (data not shown). Addition of [3-13C] serine yielded one ^{13}C-enriched atom (M+1) (data not shown). As for formate, the impact of serine on riboflavin was strongly time-dependent (Table 13). This indicated a loss or dilution of serine in the preceding growth phase.

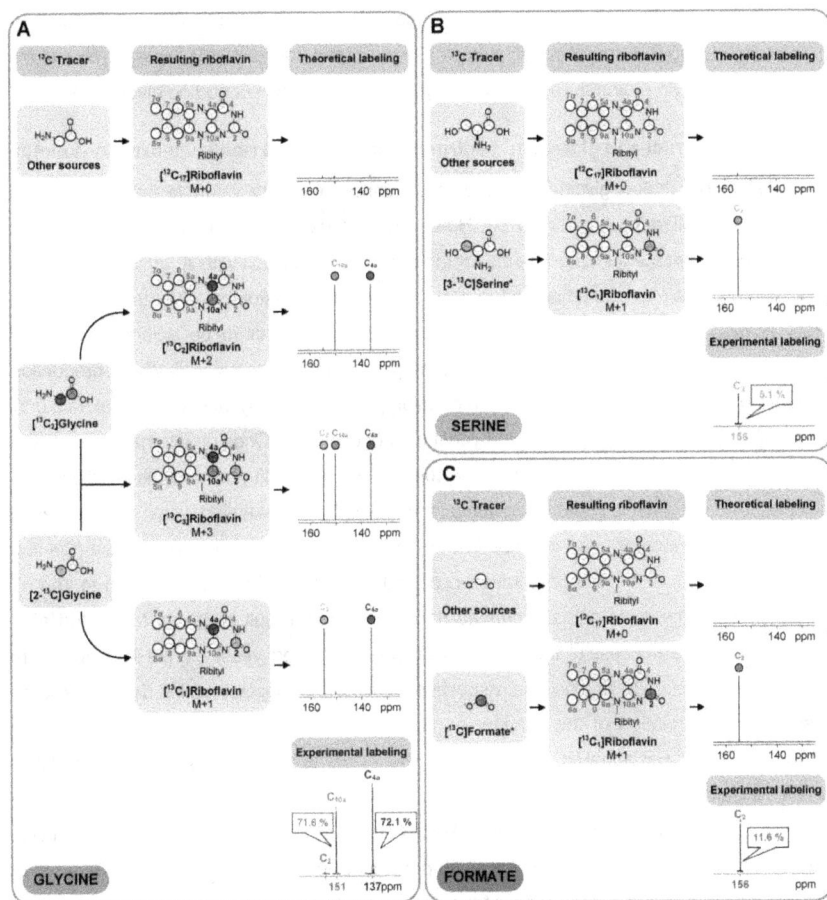

Figure 30: Theoretically possible and experimental ^{13}C NMR spectra for riboflavin produced on ^{13}C-labeled glycine (A), serine (B), and formate (C). Asterisks indicate addition of the tracer after 48 h. Glycine was added at the beginning of the cultivation. Data denote the corrected relative ^{13}C enrichment obtained from three independent replicates with a mean standard deviation of 8 %. Colored circles, ^{13}C-labeled carbon atoms; white circles, naturally labeled carbon.

For subsequent ^{13}C NMR measurements, only [3-^{13}C] serine was chosen, because the role of serine as glycine precursor had been established via GC/MS. Thus, the impact of serine as a C_1 donor was supposed to be investigated on a more molecular level. The ^{13}C NMR spectrum of riboflavin revealed a single ^{13}C relative enrichment at carbon C_2 of 5 %, the atom known to originate from the one-carbon metabolism (Figure 29E, Figure 30B, Table 14). Taken together, serine truly acted as one-carbon donor and glycine donor, but did not enter riboflavin on a different metabolic level.

Hence, for serine, a dual role was identified. In this regard, *A. gossypii* shows a similar behavior as the related fungus *Eremothecium ashbyii* (Goodwin and Jones, 1956). Here, glycine is the immediate riboflavin precursor and serine plays a central role as supporting glycine pool.

4.5.4 Yeast extract and glutamate contribute globally to the riboflavin carbon skeleton

The addition of labeled glycine, formate, and serine gave very valuable insights into the riboflavin metabolism of *A. gossypii* B2. However, rapeseed oil and yeast extract, by far the two most abundant medium ingredients for the riboflavin production with *A. gossypii* remained to be tested. Since ^{13}C-labeled rapeseed oil was unavailable, custom-synthesized fully labeled yeast extract (Chapter 4.4.4) studies were conducted. In order to be able to distinguish between incorporation of ^{13}C labeling during the growth and subsequent riboflavin production phase, two cultivations were carried out in parallel: (i) growth on fully labeled yeast extract until 32 h and (ii) growth on naturally labeled yeast extract until 32 h. Once growth had ceased and cells started to accumulate riboflavin, the cultures were centrifuged and the medium was exchanged. That way it could be distinguished, whether ^{13}C-labeled yeast extract-based compounds were incorporated into riboflavin in the early riboflavin production phase (the first 32 h of cultivation) or if the incorporation occurred after growth had ceased (32 h to 144 h). LC/MS measurements of partially purified riboflavin from the culture broth, harvested after 144 h, verified the successful incorporation of the ^{13}C-labeled yeast extract into riboflavin: 9.0 % SFL_{corr} (Table 13). The diverse respective mass isotopomer distribution did not allow the positional ^{13}C resolution of the vitamin (appendix Table 23). Therefore, at the end of the cultivation (144 h), positional ^{13}C enrichment of riboflavin was analyzed via ^{13}C NMR. The NMR results revealed that during growth, ^{13}C incorporation into riboflavin was below the detection limit and thus negligible (data not shown). Replacement of the naturally labeled medium with medium containing [U^{13}C] yeast extract (32 h post inoculation) resulted in different degrees of ^{13}C enrichment in every carbon atom of the vitamin (Table 14, Figure 31A). Obviously, yeast extract-related compounds, present in the medium at the beginning of riboflavin accumulation were used for the formation of the product. The labeling results can be lumped into three structural subunits of the vitamin: ribityl side chain, xylene ring, and pyrimidine ring. On average 8.4 % of ^{13}C labeling originating from ^{13}C-labeled yeast extract could be detected in the ribityl side chain. Surprisingly, the xylene ring, which originates from the same intracellular precursor, the pentose 5-phosphate pool (Bacher et al., 2000; Bacher et al., 1983; Bacher et al., 1985), exhibited 2.3-fold less ^{13}C incorporation (3.6 %). The pyrimidine ring, containing the four remaining carbon atoms C_2, C_4, C_{4a}, and C_{10a}, was labeled the most with 16.2 %, 12.1 %,

19.6 %, and 19.7 %, respectively. The ^{13}C labeling of C_{4a} and C_{10a} most likely originated from glycine and serine present in the yeast extract. As previously discussed, serine is converted into glycine via the SHMT (Chapter 4.4), rendering a carbon-one unit, which is then also incorporated into the vitamin at carbon atom C_2. There are two interesting details about the obtained labeling data: first, the heavy labeling of the C_4 atom, which metabolically originates from carbon dioxide fixation in the purine biosynthesis and second, the difference in labeling between ribityl side chain and xylene ring. In a previous study investigating the carbon origin of the xylene ring using precursors such as [1,3-^{13}C] or [2,3-^{13}C] ribose, labeling between those two structural groups did not differ to such great extent (Bacher et al., 1983; Bacher et al., 1985). However, in that study only the amount of *de novo* synthesized riboflavin from the respective precursor was assessed. In riboflavin, the xylene ring is the riboflavin-exclusive structural unit that contains eight carbon atoms. All other carbon is derived from the immediate riboflavin precursor GTP (Figure 32). Yeast extract is a complex nutrient, obtained from autolyzed yeast cells, which contains proteins to the largest extent, but also nucleotides or nucleobases from RNA or DNA and sugar compounds (Sørensen and Sondergaard, 2014; Zhang et al., 2003). The stronger ^{13}C labeling in the ribityl side chain compared to the xylene ring as well as the heavy labeling of the C_4 atom suggest that GTP or ATP were taken up from the yeast extract and incorporated into the vitamin. Thus, a maximum of 4 % ^{13}C enrichment, i.e. the difference in labeling of the xylene ring and the ribityl side chain, could be attributed to nucleotides in the medium originating from yeast extract. Consequently, 8 % of ^{13}C enrichment from ^{13}C-labeled yeast extract at the C_4 atom of riboflavin had to be attributed to a different origin. The ^{13}C labeling could originate from $^{13}CO_2$, which was produced during a decarboxylation reaction of a yeast extract compound. Considering the amount of ^{13}C labeling (8 %) the more likely explanation would be that guanine or adenine, also present in yeast extracts (Zhang et al., 2003), was incorporated into riboflavin via the formation of GTP (Figure 32).

Yeast extract is a complex raw material that is often employed in large-scale industrial fermentation processes of various products. However, there are large lot-to-lot variations of a single yeast extract, not to mention differences between yeast extracts produced with different strains or on different media (Diederichs et al., 2014; Sørensen and Sondergaard, 2014). The recombinant production of immunoglobulin by *S. cerevisiae* was reported to be heavily dependent on a certain yeast extract (Zhang et al., 2003). Zhang et al. (2003) especially identified adenine as a 'principle component' in the yeast extract that had a great influence on production performance of the strain used. In total, yeast extract contributed about 8 % of carbon to riboflavin. Most metabolic engineering efforts for improved riboflavin production have focused on increasing the precursor supply of glycine (Monschau et al., 1998; Schlüpen et al., 2003) and GTP (Jiménez et al., 2005; Jiménez et al., 2008; Ledesma-Amaro et al., 2015c;

Mateos et al., 2006), which are valuable components of yeast extract. In addition, the supplementation of culture media with hypoxanthine was reported to be beneficial for riboflavin production (Demain, 1972). Therefore, the yeast extract chosen for the production medium should be selected and screened carefully, since a nucleotide rich yeast extract would certainly be advantageous for an increased production performance.

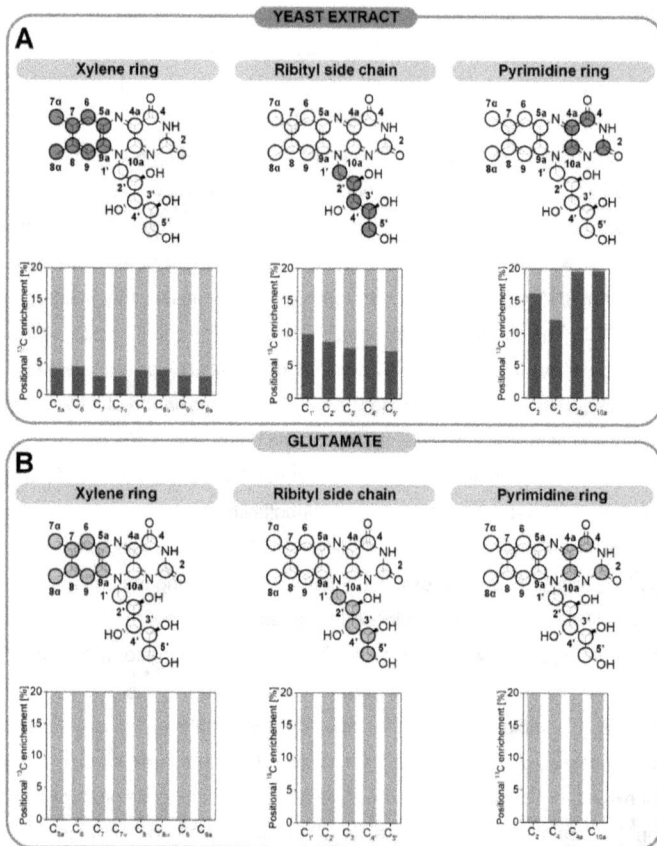

Figure 31: Experimental ^{13}C enrichment of riboflavin produced by *A. gossypii* B2 on fully ^{13}C-labeled yeast extract (A) and [^{13}C$_5$] glutamate (B) measured via ^{13}C NMR. The initial naturally labeled medium was replaced by medium containing ^{13}C-labeled yeast extract via centrifugation after 32 h of cultivation (A). The fully ^{13}C-labeled glutamate replaced the naturally labeled glutamate in the medium at the beginning of a parallel experiment with otherwise naturally labeled medium compounds (B). Riboflavin was produced on complex medium with rapeseed oil and was harvested after 144 h. The circles of the riboflavin molecules depict carbon atoms. Colored circles denote the part of the riboflavin molecule under investigation. The colors purple and light blue in the bar charts represent the percentage originating from the respective ^{13}C-labeled precursor, while the grey bars indicate origin from a different medium ingredient. Data denote values from three independent experiments with a mean standard deviation of 5 %.

Glutamate, which was also added to the medium in large amounts, was tested next as ^{13}C tracer. It replaced the naturally labeled amino acid in the initial culture medium. The obtained ^{13}C NMR data of riboflavin revealed that glutamate – similarly to yeast extract – contributed ^{13}C labeling to almost all carbon atoms in riboflavin (Table 14, Figure 31B). Glutamate is not an immediate riboflavin precursor. However, GC/MS data of cell protein revealed a close connection between α-ketoglutarate and glutamate and due to that, glutamate contributes globally to the carbon core metabolism via the TCA cycle. Hence, there is also a strong *de novo* biosynthesis of glutamate, which diluted the ^{13}C labeling of the intracellular pool. This resulted in only slight ^{13}C enrichment of the respective carbon atoms of riboflavin: 3.1 %, 1.7 %, and 0.7 % on average for ribityl side chain, xylene ring, and pyrimidine ring, respectively (Figure 31B). Surprisingly, the ribityl side chain also exhibited about 2-fold more ^{13}C enrichment than the xylene ring. Incorporation of labeled nucleotides could not explain that. So far unknown pathway activities might have resulted in the observed ^{13}C enrichment pattern.

For *B. subtilis*, the other industrially competitive riboflavin overproducer, glutamate addition has been reported for industrial media (Hohmann et al., 2011), unlike for riboflavin production with *A. gossypii*. Glutamate is metabolized by the cell and then slightly incorporated into the carbon backbone of riboflavin. Glutamate might be even more interesting when investigating the nitrogen metabolism of riboflavin: of the four nitrogen atoms, two come from glutamine, one from aspartate, and another one from glycine. Since the substitution of glutamate with glutamine led to decreased product titers (data not shown), glutamate probably resolves a potential bottleneck that cannot be eliminated by glutamine.

Riboflavin

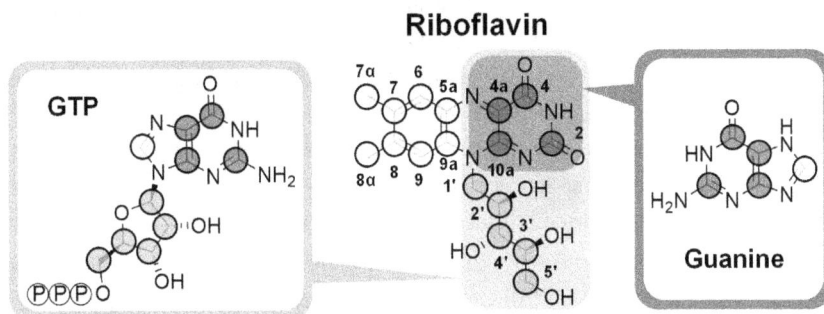

Figure 32: Structures of riboflavin, guanosine triphosphate (GTP), and guanine. GTP shares nine of its ten carbon atoms with riboflavin. Guanine shares all five carbon atoms with GTP and four of those with riboflavin. The xylene ring (white circles in riboflavin) is the structurally exclusive unit only found in the vitamin compared to the other two compounds. Grey circles denote carbon atoms that are shared between GTP, guanine, and riboflavin.

4.6 Increasing production performance via time-resolved precursor feeding

4.6.1 Extracellular formate ignites initial riboflavin overproduction

As presented in the previous chapters, the only compounds contributing to the C_1 pool in *A. gossypii* were yeast extract, serine, and formate. Glycine, however, did not function as C_1 donor. The combined contribution of the two medium ingredients yeast extract and formate to the C_2 atom in riboflavin equaled 19.9 %. This included contribution from serine present in the yeast extract, as proven by additional serine supplementation to the medium and ^{13}C NMR data of riboflavin. This left 80.1 % to another source fueling the one-carbon pool. In a tracer experiment with added [^{13}C] formate, the labeling of the extracellular formate pool was significantly diluted over time (appendix Figure 42). This indicated the presence of substantial amounts of non-labeled intracellular formate, accumulating during riboflavin production and exchanging across the cell membrane with the extracellular pool. However, even when ^{13}C NMR data of riboflavin were corrected for this dilution in input formate labeling, the combined contribution of yeast extract and extracellular formate only covered 25.8 % of the C_2 atom in riboflavin. Endogenous formate is indeed formed as a by-product in the first step of riboflavin biosynthesis (Figure 3, Figure 4), nicely matching the observations. Therefore, in the labeling experiment with [^{13}C] formate, the underlying biochemistry resulted in differently labeled intracellular and extracellular formate pools.

4.6.1.1 Intracellular formate labeling is accessible from extracellular labeling data

In order to quantify the labeling of the intracellular formate pool, extracellular ^{13}C formate and riboflavin labeling information were combined. Extracellular formate was measured via ^1H NMR as described in Chapter 3.4.4. Intracellular formate labeling had to be derived from the extracellular labeling and riboflavin labeling at carbon atom C_2 from the precursors formate and yeast extract. Following considerations regarding intracellular ^{13}C labeling of formate were made: 16.2 % ^{13}C enrichment at the C_2 of riboflavin originated from exogenous yeast extract, including serine, nucleotides, and nucleobases, taking into consideration the dilution of labeling in the culture supernatant due to naturally labeled pre-culture medium. Since serine in the yeast extract was not directly associated with riboflavin biosynthesis, intra- and extracellular labeling were assumed to be equilibrated. This left 83.8 % of positional enrichment to formate, whether endogenous or exogenous. The percentage of labeling at this carbon atom in riboflavin derived from extracellular formate added at 0 h was 3.7 % without taking into account dilution effects throughout the cultivation. Dilution of extracellular formate labeling started at the beginning of riboflavin biosynthesis as it is directly linked to it. Thus, with increasing riboflavin concentration, the amount of naturally labeled formate would increase and the degree of ^{13}C dilution would increase in return. In order to now derive intracellular formate labeling from extracellular formate labeling a linear correction function was applied. Assuming that during growth, formate production from unlabeled substrates was negligible (correction factor 1) and riboflavin biosynthesis is linear, the degree of dilution of labeling increased until the end of the cultivation (Figure 33). The value for the maximum correction factor was set as ratio between theoretical labeling from formate (83.8 %) and the experimental labeling from exogenous formate (3.7 %): 22.6.

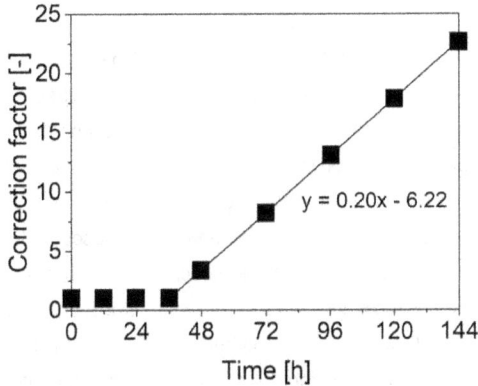

Figure 33: Function for the correction factor for estimation of intracellular ^{13}C formate labeling from extracellular ^{13}C enrichment of formate. Correction factor is constant during growth as it is assumed that only exogenous formate is fed into the metabolism. Once riboflavin production starts (after 36 h), formate synthesis from ribulose 5-phosphate begins in a linear manner giving rise to a linearly increasing correction factor. Depending on the point in time, extracellular ^{13}C formate labeling is then divided by the corresponding factor. These data are then used in the model simulating the transmembrane flux (Figure 12, appendix Chapter 6.9).

4.6.1.2 Formate labeling allows for transmembrane flux simulations

Using a computational model, the data were used to simulate the transmembrane exchange flux of formate for different intervals during riboflavin biosynthesis (Figure 34). Formate was added at the beginning of the cultivation. During the first 36 h formate flux was directed into the cell with a net flux of 87 to 83 % (Figure 34A and Figure 34B). Nearly no formate was produced from the pentose 5-phosphate pool, since almost no riboflavin was formed during that period. The small amount of riboflavin that was synthesized was fed from a formate pool, which almost exclusively originated from exogenous formate. Within the next 12 h, more and more formate was released from the intracellular unlabeled carbon pool, giving rise to an increase in transmembrane formate exchange. As riboflavin biosynthesis progressed, formate derived from unlabeled carbon gained increasing importance, which is illustrated by the rising flux from the pentose 5-phosphate pool (583 % for 48 to 72 h and 1040 % for 72 to 96 h, Figure

34D and Figure 34E). Ultimately, this large flux from the unlabeled intracellular formate pool resulted in a net efflux of 334 % and subsequent accumulation of formate in the culture supernatant, which was confirmed by a final formate titer of 0.46 g L^{-1} in the culture broth (data not shown). The flux data showed that the initial biosynthetic phase of riboflavin heavily depends on exogenous formate. Availability of formate seemed a critical step in the early biosynthesis, however, during the later phase riboflavin production became independent from extracellular formate.

As described for yeast, mainly *S. cerevisiae*, there are three donors for the folate-dependent C_1 pool: formate, serine, and glycine (Piper et al., 2000), which enter C_1 metabolism at different oxidation levels. In the set-up presented here, only yeast extract – containing serine, nucleotides, and nucleobases – and formate donated to one-carbon units in the cell and to the carbon atom C_2 of riboflavin. Extracellular formate contributed 4 % to the C_2 atom of riboflavin (Figure 30C, Table 14), whereas yeast extract contributed 16 % (Figure 31A, Table 14), 12 % of which probably stemmed from nucleotides and nucleobases with the remaining 4 % originating from serine. The inspection of all data, including the surprisingly heavy dilution of the ^{13}C labeling of formate in the medium (appendix Figure 42), suggested that the intracellular formate pool is far less ^{13}C-labeled and contributes the remaining 80 % of the C_2 atom in riboflavin. Where does the non-labeled formate come from? Net formate synthesis is directly associated with riboflavin biosynthesis: ribulose 5-phosphate is converted to DHBP (Figure 3, Figure 4), which releases one formate, originating from the sugar moiety (Bacher et al., 1998). A second GTP-derived formate, transiently released during riboflavin biosynthesis (Bacher et al., 2000), already stems from the intracellular 10-CHO-THF pool and does not contribute likewise. In the given tracer set-up, the on-going riboflavin biosynthesis, thus, inherently accumulates increasing amounts of non-labeled formate, matching the NMR data. On this basis, ultimately yeast extract and formate remained as exclusive carbon-one donors. Simulation of transmembrane fluxes of formate unraveled a crucial contribution of exogenous formate availability to the riboflavin biosynthesis. In the early phase of production, extracellular formate displayed the dominant source of the carbon-one pool. The direction of flux was from the medium into the cell. In later phases, the biosynthetic pathway more and more provided its own one-carbon unit, which further supported riboflavin biosynthesis in an autocatalytic manner. Towards the end of the cultivation, the flux was reverted and exported excess formate. In *S. cerevisiae* formate transport across the membrane was described as proton-symport, which requires energy, as well as transporter independent diffusion of the protonated acid (Casal et al., 1996; Geertman et al., 2006). Most important, the obvious impact on the early production phase suggested to add formate to the medium formulation.

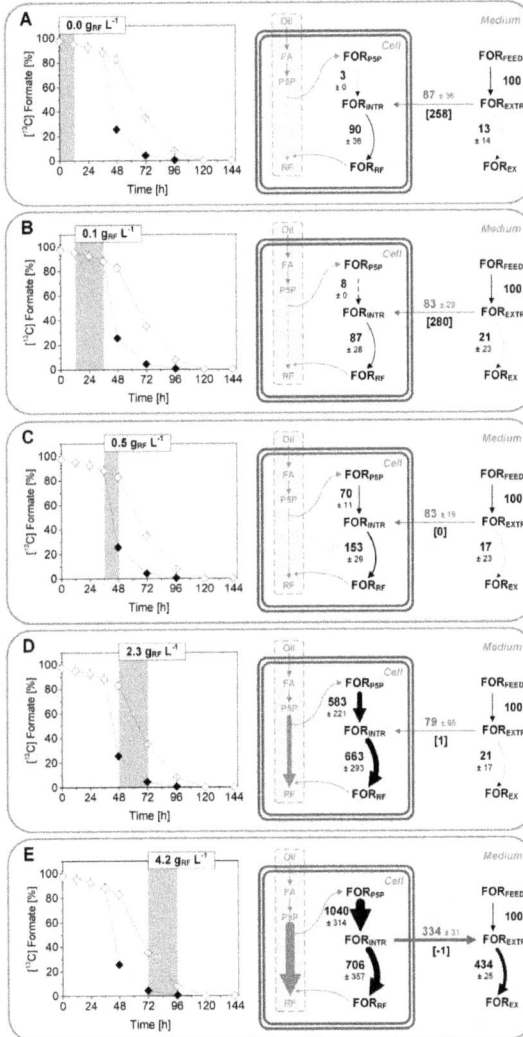

Figure 34: Transmembrane formate flux at different states of riboflavin production by *A. gossypii* B2. Cultures were grown on complex medium with [^{13}C] formate. The ^{13}C enrichment of extracellular formate (white diamonds) was measured using ^{1}H NMR. Intracellular formate labeling (black diamonds) was derived from ^{1}H NMR data of exogenous formate and ^{13}C NMR data of riboflavin. Red arrows indicate the net transmembrane flux. Values in brackets indicate degree of reversibility. All fluxes are expressed as molar percentages of the corresponding specific formate uptake rate represented by the net flux of formate across the plasma membrane (A: 71.6 µmol g^{-1} h^{-1}, B: 135.0 µmol g^{-1} h^{-1}, C: 9.1 µmol g^{-1} h^{-1}, D: 0.1 µmol g^{-1} h^{-1}, E: -0.9 µmol g^{-1} h^{-1},). The errors represent 90 % confidence intervals and were calculated by Monte-Carlo analysis. Gray boxes depict the time interval for the corresponding transmembrane flux distribution. All fluxes are given in percent. FA, fatty acids; FOR$_{EX}$, formate pool for metabolite balancing; FOR$_{EXTR}$, extracellular formate pool; FOR$_{PSP}$ formate pool from ribulose 5-phosphate; FOR$_{RF}$, formate pool entering riboflavin as C$_1$ unit; RF, riboflavin.

93

At this stage, it cannot be excluded that formate, beyond being a key carbon-one donor for riboflavin, also causes other, more indirect effects that stimulate product formation. As example, formate was reported to influence phosphorylation of the formate dehydrogenase (FDH) in yeast (Bykova et al., 2003). Further work will be needed to completely understand its role. Its impact, however, is without doubt.

4.6.2 The C_1 metabolism displays a bottleneck of initial riboflavin overproduction

4.6.2.1 Formate specifically added during initial riboflavin production increased performance

The flux simulations and labeling data from the previous chapters showed that extracellular formate has an impact on riboflavin biosynthesis, which is strongly time-dependent. Thus, the effect of time-resolved formate addition was assessed for that specific time frame in a series of experiments. Formate (2 g L^{-1}, 30 mM) was added to the medium not at the beginning of cultivation, but after 12 h, 24 h, and 36 h, respectively (Figure 35A). The standard cultivation with formate in the initial medium served as control. Formate addition after 12 h resulted in a final riboflavin titer increase of 45 %. Feeding formate at a later point in time did not increase riboflavin titer significantly. These results underlined the formate flux data: availability of exogenous formate is important during an early phase of cultivation and is able to increase riboflavin concentration drastically, when fed at a very specific point during the cultivation. The formate concentration applied in this study was not found inhibitory, but growth was unaffected. Even when 60 mM formate (4 g L^{-1}) were added (data not shown), growth was not impaired.

4.6.2.2 Serine supplementation supports riboflavin production

Data presented in the previous chapters demonstrated the dual role of serine: a one-carbon as well as a glycine donor. Replacement of glycine by serine led to a decreased final riboflavin concentration (data not shown) and was not profitable for the process. Here, the capacities of serine as C_1 donor were investigated. Since the targeted formate addition successfully increased productivity of A. gossypii B2, the impact of serine was investigated in a similar manner. For that, formate was substituted by serine in the medium and serine was added in a time-resolved mode. Unlike formate, addition of serine at the very beginning of cultivation (0 h) was very effective as the final riboflavin titer could be increased again by 45 % compared to the standard cultivation (Figure 35B). Addition of serine after 12 h and 24 h was still beneficial with an increase of 33 % each. Regarding positive effects of formate and serine on riboflavin biosynthesis, both compounds shared the same trend: addition in the early cultivation phase was advantageous. Formate exhibited a narrow time optimum, whereas the positive influence of serine showed a broader range between 0 to 24 h. Selected experiments revealed that this type of formate and serine supplementation was also advantageous for the *A. gossypii* WT (data not shown). Since the time specific addition of each of the C_1 donors resulted in such a tremendous increase in riboflavin titer, the combined addition of the two donors was tested. For a medium, which contained formate, serine addition was most advantageous, when added after 36 h (data not shown). Thus, the final combinatorial set-up was cultivation on a medium with neither formate nor serine. After 12 h formate was added, followed by serine supplementation after 36 h. This fine-tuned approach resulted in a final product titer increase of 56 % (data not shown). While this set-up boosted the production performance of the overproducing strain by another 11 %, feasibility of an industrial process relies on profit margins and the use of an additional medium component must be carefully considered with regard to substrate costs and product price.

4.6.2.3 Improved performance relies on intracellular precursor availability

As shown above, targeted addition of formate after 12 h of cultivation and substitution of formate with serine in the initial culture medium both led to improved product formation (Figure 35). In order to gain a deeper understanding of the underlying mechanisms, intracellular concentrations of amino acids and formate were measured. *A. gossypii* was grown under three different conditions. This included a control (addition of formate to the initial medium) and the two identified optimal conditions: formate addition after 12 h, and substitution of initial formate with serine. Samples were taken throughout the riboflavin biosynthetic phase from 48 to 120 h and analyzed for the intracellular levels of the riboflavin precursors: glycine, serine, and formate (Figure 35C-E). Both improved production conditions exhibited a strongly enhanced intracellular formate pool after 48 h: 251 % and 120 % increase, respectively. Since the intracellular formate level increased throughout riboflavin production in the control condition, pool sizes of the different conditions were converging toward the later riboflavin phase (Figure 35C). Regarding amino acid concentrations, addition of formate after 12 h had the most pronounced effects compared to the control. After 72 h, glycine and alanine concentrations were increased by 77.5 % and 65.7 %, respectively (Figure 35D and F). The difference in serine was even more remarkable: 167.8 µmol g_{CDW}^{-1} (control: 37.8 µmol g_{CDW}^{-1}) (Figure 35E). Substitution of formate by serine only led to a few significant changes compared to the control: intracellular serine was 4.2-fold increased after 48 h with regards to the control (Figure 35E). In addition, glutamate and alanine concentrations were elevated after 48 h with 144.9 µmol g_{CDW}^{-1} and 15.0 µmol g_{CDW}^{-1} (control: 112.3 µmol g_{CDW}^{-1} and 12.2 µmol g_{CDW}^{-1}, respectively) (Figure 35F and G). Thus, intracellular availability of the three precursors (formate, serine, and glycine) played a crucial role in improved performance of riboflavin production.

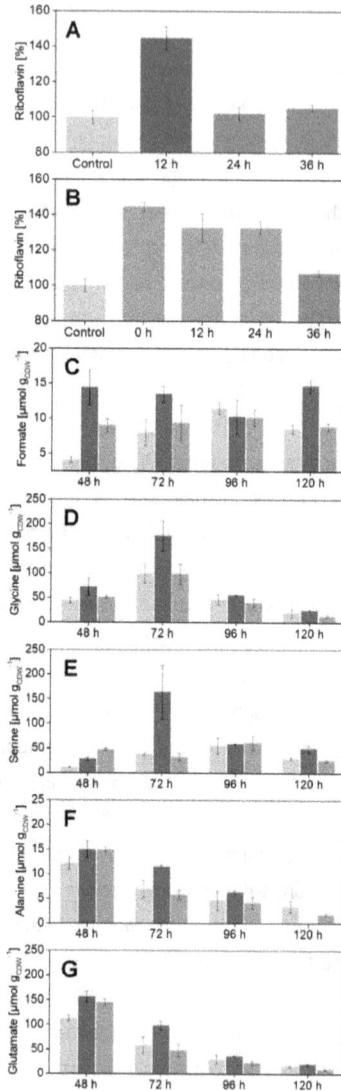

Figure 35: Time-resolved addition of formate (A) and serine (B) to *A. gossypii* B2 cultures led to increased product titers. Formate was added at various points in time in parallel experiments and riboflavin titer was determined after 144 h. Formate added at 0 h served as control (light grey). Serine replaced formate in the medium and was added after different times during cultivation. Riboflavin titer was determined after 144 h and a culture with formate added at 0 h instead of serine was used as control (light grey). Intracellular metabolites were sampled at various points in time and were measured via ¹H NMR (C) or HPLC (D-G) for three different conditions: control condition with formate added at 0 h (light grey), formate addition after 12 h (red), and substitution of formate with serine at 0 h (green).

Comparison of intracellular amino acid levels with different yeasts shows that while the glutamate pool is large in all strains, increased intracellular glycine and serine concentrations, with one exception, are unique to *A. gossypii* (Figure 36). Compared to *S. cerevisiae*, glycine is increased 3-fold (*A. gossypii*c) or even 5-fold (*A. gossypii*F12) (Bolten and Wittmann, 2008).This emphasizes the importance of elevated intracellular glycine and serine pools for riboflavin overproduction.

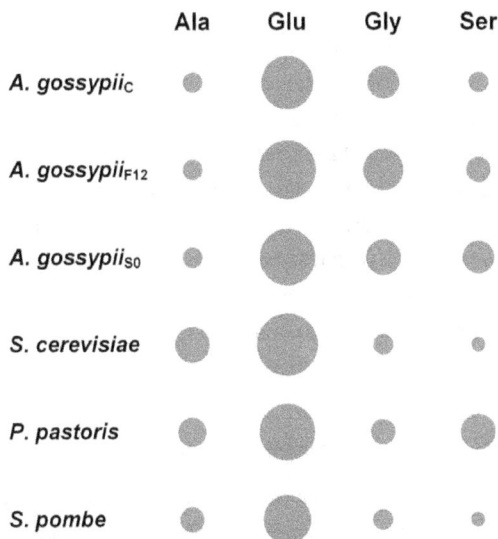

Figure 36: Comparison of intracellular amino acid concentrations in various yeast strains. Size of circle correlates with amino acid concentration in μmol g_{CDW}^{-1}. *A. gossypii* was grown on complex medium and vegetable oil as described above. Samples were taken after 48 h. The yeast strains were cultivated on minimal medium and glucose, which was supplemented with yeast extract and tryptone for all strains, but *S. cerevisiae*. *A. gossypii*c, *A. gossypii* cultivated under standard conditions; *A. gossypii*F12, *A. gossypii* cultivated on medium with formate addition after 12 h; *A. gossypii*S0, *A. gossypii* cultivated on medium with substitution of initial formate by serine; *S. cerevisiae*, *Saccharomyces cerevisiae*; *P. pastoris*, *Pichia pastoris*; *S. pombe*, *Schizosaccharomyces pombe* (Bolten and Wittmann, 2008).

Based on the ^{13}C labeling data, formate and serine were identified as donors of the carbon-one pool. Moreover, the studied producing strain made efficient use of extracellular formate during the early production phase, which indicated a potential limitation of the carbon-one pool inside the cells. These observations displayed a rational basis to improve the formation of the vitamin: the extra supply of carbon-one precursors in a way that enhances their intracellular availability specifically at the time point, when they were needed most. The time-resolved addition of formate obviously was a smart and easy way to reach that. Supplementation with about 2 g L^{-1} (30 mM) formate after 12 h boosted the riboflavin titer by 45 % (Figure 35A), whereas initial and also later addition did not cause significant changes, nicely matching the ignition hypothesis. In a similar manner, also replacement of the initial formate by serine had a positive effect and led to substantial titer increase. The specific addition of the two nutrients resulted in increased intracellular pools of all three key precursors, i.e. glycine, serine, and formate, which can be regarded as highly beneficial for production performance (Figure 35). So far, both compounds have not been considered as part of production media, but glycine has largely been the major supplement (Sahm et al., 2013). There are different reports on the toxicity of formate in fungi or yeast. The minimum inhibitory concentrations vary depending on the process or the strain under investigation (Babel et al., 1983; Du et al., 2008; Lastauskienė et al., 2014). Here, formate was not found inhibitory. Hereby, the understanding of the molecular details required to add only a small amount at one specific time point, which was then quickly consumed. This might have been beneficial to avoid toxic effects. For industrial production, time-resolved supplementation seems easily applicable. Considering that riboflavin production with *A. gossypii* is already operated in fed-batch mode with a preceding batch phase (Schwechheimer et al., 2016), delayed formate addition appears straightforward. Addition of serine to the initial production medium could be realized instantly, however, economic feasibility of such a process also needs to be considered with respect to substrate availability and profit margins.

4.7 The complete picture: riboflavin biosynthesis on vegetable oil

Resolving the metabolic fluxes from various potential building blocks of a complex medium towards riboflavin biosynthesis is a challenging task. Pioneering labeling studies have focused on the addition of single tracer compounds to the cultivation, which were mostly done using glucose or other sugars as main carbon source (Bacher et al., 1998; Jeong et al., 2015; Plaut, 1954a; Plaut, 1954b; Plaut and Broberg, 1956; Schlüpen et al., 2003). While this early work delivered important details about the biosynthetic pathway, the full picture was still missing. In this work, ^{13}C labeling studies were embedded into a sophisticated framework of experimental design, multi-readout ^{13}C labeling analysis by a combination of different MS and NMR techniques, and model-based data processing. This facilitated the use of a complex medium and rapeseed oil as a carbon source, which resembled industrial process conditions. The integrated rich data set now enabled the calculation of a metabolic flux map, comprising the respective contributions of single medium ingredients to riboflavin in this environment (Figure 37). A specific approach for data processing and carbon balancing was chosen to finally derive the flux distribution. Different from conventional flux studies on only one substrate, with precise molecular fluxes (Kohlstedt et al., 2014; Lange et al., 2017), this work had to consider the positional contribution of individual precursors to specific parts or even single carbon atoms of the riboflavin molecule. That is, why the flux estimation here was taken to the level of single carbon atoms, and fluxes were split into fluxes from individual carbon atoms of nutrients, which converged within the metabolism to form molecules or parts of the vitamin. This perfectly matched the type of labeling information, i.e. positional ^{13}C enrichment that was available to great detail from the conducted labeling studies. The positional ^{13}C enrichments of the single tracer experiments (Table 14) were integrated with the pathway stoichiometry. In addition, carbon balancing was applied. The assumption of steady-state, required for this approach, appeared justified, considering that the change of intracellular metabolite pools was small compared to fluxes through those pools, similar to a previous approach (Krömer et al., 2004). Together this rendered a flux distribution based on single carbon atoms (Figure 37). Therefore, the flux into a single carbon atom was set to 1. Likewise, the flux into riboflavin equaled 17, since the vitamin contains seventeen carbon atoms and the flux of e.g. GTP, a ten-carbon molecule, toward riboflavin biosynthesis corresponded to the value 10. Accordingly, fluxes between other metabolites were based on the sum of the fluxes of their carbon atoms and additionally, considering the ^{13}C labeling from the respective ^{13}C tracers. For each carbon atom of riboflavin, the metabolic precursor was determined (Figure 45, Figure 46, Figure 47). The way, in which the individual medium ingredients, e.g. yeast extract, would contribute to that metabolic precursor was evaluated and for each case, the entry point into the metabolism was chosen based on literature (Bacher et al., 1998; Plaut, 1954a; Plaut, 1954b; Plaut and Broberg, 1956) as well as data obtained in this study (Table 14). As an example, the C_{4a} atom of riboflavin is derived from intracellular glycine (Figure 45) (Plaut, 1954a). In the presented set-up, glycine could originate from extracellular glycine, which was supplemented to the medium, glycine from yeast extract, but also serine from yeast extract, adenine/guanine and ATP/GTP from yeast extract or from rapeseed oil, via 3-phosphoglycerate and serine (appendix Chapter 6.10, Figure 45). Thus, the glycine unit of riboflavin could be derived from three different medium compounds, i.e. glycine, vegetable oil, and yeast extract, and four different metabolites, i.e. glycine, serine, adenine/guanine, or ATP/GTP. Adenine/guanine as well as ATP/GTP are treated as one group, respectively, since the data did not allow a distinction between the individual compounds. Analogous considerations were made for every carbon atom and fluxes were calculated accordingly (Chapter 6.10).

The full picture of carbon contribution to riboflavin in its biosynthetic phase is depicted in Figure 37. It is important to note that only the carbon assimilation reactions directly fueling riboflavin biosynthesis are considered for this approach. While the single carbon fluxes in Figure 37 are expressed as single carbon atoms based on calculations as described in the appendix (Chapter 6.10), these fluxes were then converted into molar fluxes: first, each flux was divided by the number of carbon atoms of the respective metabolite, then, the flux into riboflavin was set to 100 % and represented the specific riboflavin production rate (8.7 µmol g^{-1} h^{-1} ± 2.3 µmol g^{-1} h^{-1}). Subsequently, all other fluxes were normalized to that flux into riboflavin. The resulting relative carbon fluxes are given in Figure 38. Rapeseed oil was cleaved via the extracellular lipase into one mole glycerol and three moles fatty acids (Stahmann et al., 1994). A high flux through the ß-oxidation pathway (1160 %) and the glyoxylate shunt (580 %) could be observed. Acetyl-CoA generated in the ß-oxidation pathway was then converted into malate through the glyoxylate shunt in order to enter gluconeogenesis and subsequently serve as carbon donor from vegetable oil for riboflavin. Therefore, all carbon in riboflavin that was derived from rapeseed oil, was assimilated via the glyoxylate shunt. Glutamate most likely entered the metabolism through α-ketoglutarate with a flux of 12 %, which resulted in a comparably low flux from α-ketoglutarate to oxaloacetate (12 %). A flux of 592 % was directed through the lower gluconeogenic pathway. Here, the map representing single carbon atoms (Figure 37), highlights the loss of carbon (5.9) via decarboxylation of the OAA/malate pool to the PEP/pyruvate pool. The assembly of two three-carbon units to one six-carbon molecule resulted in a decreased flux (288 %) in the upper gluconeogenesis (Figure 38). Another decarboxylation reaction resulted in loss of carbon upon entry into the PP pathway (2.88) (Figure 37). The carbon flux of 288 % was split at ribulose 5-phosphate, an immediate precursor of riboflavin. The exact pathway used to synthesize this pool, gluconeogenesis or PP pathway, could not be distinguished by this approach. The combined carbon flux from ribulose 5-phosphate pool, which contributes eight carbon atoms in total to riboflavin, towards riboflavin was 200 %, with a corresponding carbon flux towards the intracellular formate pool of 100 % each. The second branch from ribulose 5-phosphate was directed into purine biosynthesis (96 %). The flux through the complete *de novo* biosynthesis of purines (GTP pool) equaled 88 %. Throughout that pathway, one-carbon units as well as carbon dioxide and glycine were incorporated into the final GTP molecule. Nucleotides from yeast extract (ATP or GTP) donated another 4 % to that pool. However, ^{13}C labeling data also suggested that adenine or guanine from the supplemented yeast extract together with ribose 5-phosphate contributed 8 % to the GTP pool of riboflavin (Figure 38). The carbon flux distribution highlights the dominant role of vegetable oil in riboflavin biosynthesis. Indeed, the overall flux of rapeseed oil into riboflavin equaled 13.8 carbon atoms (81 % of the molecule). However, the contribution of other medium ingredients to riboflavin, especially to the pyrimidine ring, was as high as 19 %. This especially included glycine and C_1 donors. As shown in the previous chapter, overcoming a limitation in the one-carbon pool, which makes up only 6 % (1 carbon atom) of the vitamin, can have a great impact on product titers. Consequently, the knowledge gained in such ^{13}C tracer studies conveys a valuable starting point for rational strain engineering as well as carefully designed process control.

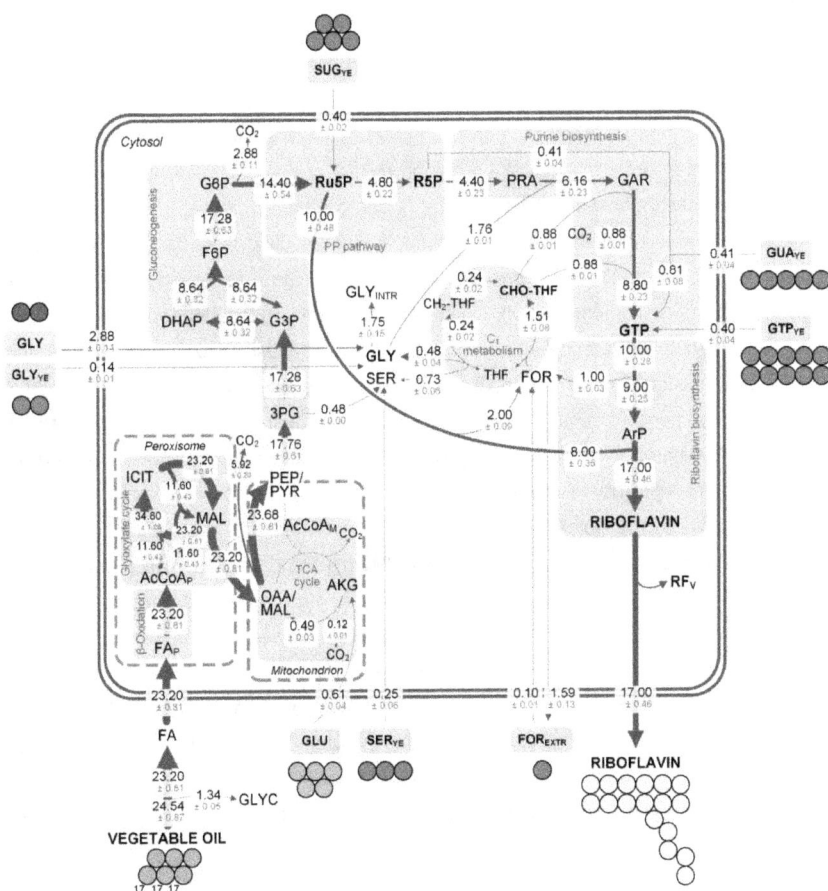

Figure 37: Carbon contribution of medium ingredients to riboflavin based on combined parallel ^{13}C-labeled tracer studies using [[13]C2] glycine, [^{13}C] formate, [[13]C5] glutamate, and [U[13]C] yeast extract. Riboflavin was synthesized by *A. gossypii* B2 grown on complex medium with rapeseed oil. Riboflavin was obtained at the end of the growth phase of riboflavin producing *A. gossypii* after 144 h. Data are derived from positional ^{13}C enrichment obtained from ^{13}C NMR measurements (Table 14), corrected for natural labeling and dilution effects through unlabeled pre-culture medium. Values denote carbon atoms and are normalized to 17 carbon atom influx into riboflavin. All values were multiplied by the number of carbon atoms of the reactants. The arrow thickness is proportional to the corresponding flux. The direction of net fluxes is indicated by size of arrow head. Note that the model is simplified and cannot distinguish between carbon flux through e.g. gluconeogenesis or lower PP pathway as well as pyruvate dehydrogenase. Reaction between OAA/MAL and PEP/PYR pool is a lumped flux. Only reactions necessary for riboflavin biosynthesis were considered and all reactions represent net fluxes. Note that the conversion of citrate to isocitrate via aconitase most likely does not occur in the peroxisome (Murakami and Yoshino, 1997). 3PG, 3-phosphoglycerate; AcCoA$_{P/M}$, peroxisomal/mitochondrial acetyl-CoA; AKG, α-ketoglutarate; ArP, 5-amino-6-ribitylamino-2,4(1H,3H)-pyrimidinedione; CH2-THF, 5,10-methylenetetrahydrofolate; CHO-THF, 10-formyltetrahydrofolate; FA, fatty acids (here: three C17.3 FA); FOR, formate; GAR, glycineamide ribonucleotide; GLU, glutamate; GLY, glycine; GLY$_{INTR}$, intracellular glycine pool; GTP, guanosine triphosphate; PRA, 5-phosphoribosylamine; PYR, pyruvate; R5P, ribose 5-phosphate; Ru5P, ribulose 5-phosphate; RF$_V$, riboflavin stored in the vacuole; SER, serine; THF, tetrahydrofolate; YE, yeast extract.

101

The fluxes presented here (Figure 37, Figure 38), considered riboflavin biosynthetic reactions and were based on ^{13}C labeling information obtained through ^{13}C NMR analyses. All other reactions regarding the metabolism such as maintenance could not be captured. Keeping that in mind, the fact that 23.2 carbon atoms from vegetable oil have to be assimilated in the glyoxylate shunt in order to obtain 14.4 carbon atoms for riboflavin biosynthesis in the PP pathway illustrates the extent to which carbon is lost to carbon dioxide (Figure 37). This can be expressed by the biosynthetic efficiency of riboflavin production under the given conditions, which is based on stoichiometry and equals 73 % with regard to rapeseed oil and 60 % regarding all carbon sources, i.e. oil, yeast extract, glycine, glutamate, and formate. Thus, 73 % of all oil-derived carbon needed for riboflavin biosynthesis actually ended up in the vitamin itself, the rest mainly lost to decarboxylation. The basic biosynthetic efficiency for riboflavin biosynthesis from vegetable oil is 61 %, which means that all carbon needed for riboflavin biosynthesis is derived from vegetable oil and riboflavin is, therefore, fully *de novo* synthesized. The efficiency calculated for riboflavin with regard to oil for the set-up presented here is increased by the factor 1.2, therefore indicating that not only rapeseed oil is used for the formation of the vitamin. This was obviously shown through the combination of extensive ^{13}C tracer experiments. The maximum biosynthetic efficiency, i.e. the assembly of riboflavin from the medium-derived compounds GTP and pentoses, equals 85 %. Even though, yeast-extract based substrates as well as supplemented glycine were also incorporated into riboflavin under the given conditions, the overall biosynthetic efficiency is minimal (60 %). This is in part due to decarboxylation, however, also the inefficient use of glycine (potential storage in vacuole or degradation in the metabolism) (Messenguy et al., 1980; Schlüpen et al., 2003) as well as production of formate during riboflavin biosynthesis as previously shown (Chapter 4.6) result in loss of carbon.

The overall yield for riboflavin biosynthesis on oil was 0.28 $mol_{RF}\ mol_{Oil}^{-1}$ (Table 6). The conversion of the molar yield into a C-molar yield rendered 0.09 $C\text{-}mol_{RF}\ C\text{-}mol_{Oil}^{-1}$, i.e. 9 %. Therefore, 91 % of carbon from rapeseed oil, which was taken up by the cell, did not contribute to riboflavin biosynthesis. Cells were stationary during riboflavin biosynthesis (Figure 13A) and conversely, significant growth or accumulation of intracellular storage lipids could also be ruled out as carbon consuming reaction. Since no significant by-products could be measured, the carbon that was still taken up during the production phase, i.e. fatty acids and glutamate amongst others (Figure 13), was likely also decarboxylated via a highly active TCA cycle, as shown for the growth phase of A. gossypii on oil (Figure 27). Yet, the question remains, why the TCA cycle is so active. During growth, most cellular building blocks were taken up from the medium, thus resulting in a low ATP demand (Figure 27). During riboflavin biosynthesis, 9-fold more oil-based carbon was taken up than was used for the product. Little is known about the

TCA cycle activity of riboflavin producing *A. gossypii* on vegetable oil or pathway activity of other microorganisms on oil as a substrate. It was shown that *A. gossypii* accumulates intracellular *trans*-aconitate during growth (Sugimoto et al., 2014). This known inhibitor of the enzyme aconitase, which catalyzes the conversion of citrate to isocitrate via the intermediate *cis*-aconitate, is apparently formed spontaneously from *cis*-aconitate (Sugimoto et al., 2014). During the riboflavin production phase, expression levels of another enzyme, *trans*-aconitate 3-methyltransferase, were increased nearly 3-fold compared to the growth phase. Through methylation of *trans*-aconitate, this enzyme relieved the inhibition of the TCA cycle enzyme aconitase and consequently, enabled an increased flux through the TCA cycle as well as the glyoxylate shunt (Sugimoto et al., 2014). In a study with mice that were put on a high-fat diet, liver metabolism was investigated (Satapati et al., 2012). The consequences of the fatty diet were increasing hepatic insulin resistance due to the lipid overload, but also an increased TCA cycle as well as elevated gluconeogenesis. The increased flux through the TCA cycle possibly compensated for an inefficiently coupled oxygen consumption and ATP synthesis, i.e. a dysfunctional mitochondrial respiration. The authors also observed the production of reactive oxygen species, thereby linking the induction of a highly active TCA cycle to oxidative stress (Satapati et al., 2012). Indeed, mitochondria produce reactive oxygen species when e.g. the NADH pool is reduced as a result of low ATP demand (Murphy, 2009). This is especially interesting, because riboflavin biosynthesis in *A. gossypii* is associated with oxidative stress (Kavitha and Chandra, 2009; Kavitha and Chandra, 2014). These studies provide evidence for an increased TCA cycle flux during growth on lipid-rich media, which might also apply for *A. gossypii*. However, while these connections offer interesting starting points for a deeper assessment of the underlying mechanisms, further work will undoubtedly be needed to resolve the fine structures of riboflavin biosynthesis with *A. gossypii* on rapeseed oil.

As shown, the combined analysis of parallel ^{13}C tracer experiment allowed the calculation of a carbon flux distribution for riboflavin producing *A. gossypii*. In a complex cultivation set-up, containing rapeseed oil and yeast extract, in addition to other supplements, the contribution of each medium ingredient to riboflavin could be quantified precisely. The data showed that while yeast extract was the main carbon source for growth (Figure 23), its impact on riboflavin biosynthesis was still significant with an overall contribution of 8 %. In addition, glycine as two-carbon unit, which is readily incorporated into the vitamin, also made up 8.5 % of the product. By far the greatest impact had rapeseed oil (81 %), which is also described as superior carbon source compared to glucose (Demain, 1972).

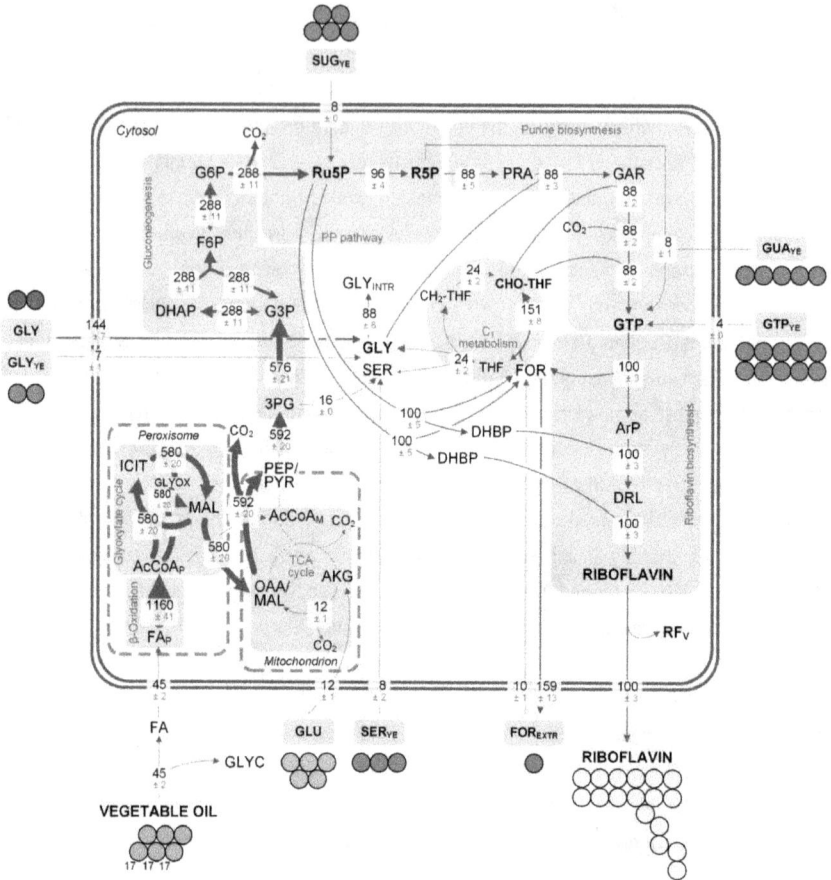

Figure 38: Intracellular carbon fluxes of riboflavin production with *A. gossypii* B2 on rapeseed oil and complex medium, which were determined by four parallel ^{13}C-labeled tracer studies using [^{13}C$_2$] glycine, [^{13}C] formate, [^{13}C$_5$] glutamate, and [U^{13}C] yeast extract. The carbon fluxes were normalized to the specific riboflavin production rate (8.7 µmol g^{-1} h^{-1} ± 2.3 µmol g^{-1} h^{-1}), which was set to 100 %. Riboflavin was obtained at the end of the growth phase of riboflavin producing *A. gossypii* after 144 h. Data are derived from positional ^{13}C enrichment obtained from ^{13}C NMR measurements (Table 14), corrected for natural labeling and dilution effects through unlabeled pre-culture medium. Note that the model is simplified and cannot distinguish between carbon flux through e.g. gluconeogenesis or lower PP pathway as well as pyruvate dehydrogenase. The arrow thickness is proportional to the corresponding flux. The direction of net fluxes is indicated by size of arrow head. Reaction between OAA/MAL and PEP/PYR pool is a lumped flux. Only reactions necessary for riboflavin biosynthesis were considered and all reactions represent net fluxes. Note that some reactions as shown in Figure 37 can be expressed as a single reaction, i.e. all decarboxylation reactions and in general all reactions with more than one educt or product can be expressed as a single flux. All fluxes from the medium into the cell are not fluxes based on concentrations, but solely derived from ^{13}C labeling of riboflavin. Note that the conversion of citrate to isocitrate via aconitase most likely does not occur in the peroxisome (Murakami and Yoshino, 1997). 3PG, 3-phosphoplycerate; AcCoA$_{P/M}$, peroxisomal/mitochondrial acetyl-CoA; AKG, α-ketoglutarate; ArP, 5-amino-6-ribitylamino-2,4(1H,3H)-pyrimidinedione; DRL, 6,7-dimethyl-8-ribityllumazine; CH$_2$-THF, 5,10-methylenetetrahydrofolate; CHO-THF, 10-formyltetrahydrofolate; FA, fatty acids (here: three C17.3 FA); FOR, formate; GAR, glycineamide ribonucleotide; GLU, glutamate; GLY, glycine; GLY$_{INTR}$, intracellular glycine pool; GTP, guanosine triphosphate; PRA, 5-phosphoribosylamine; PYR, pyruvate; R5P, ribose 5-phosphate; Ru5P, ribulose 5-phosphate; RF$_V$, riboflavin stored in the vacuole; SER, serine; THF, tetrahydrofolate; YE, yeast extract.

5 CONCLUSION AND OUTLOOK

The industrial production of riboflavin is one of the great success stories in biotechnology and metabolic engineering (Revuelta et al., 2016). Within just a few years, fermentative synthesis completely replaced the more than 50-year old chemical synthesis of this vitamin (Schwechheimer et al., 2016). However, in order to keep up with constantly increasing market demands, continuous improvement of production strains and process conditions is fundamental.

In this work, growth and production behavior of riboflavin producing wild type *A. gossypii* WT and overproducing mutant *A. gossypii* B2 were investigated on different substrates with the combined use of GC/MS, LC/MS as well as NMR. Pioneering work on *A. gossypii* and riboflavin metabolism were only feasible on glucose or ethanol as main carbon source (Bacher et al., 1998; de Graaf et al., 2000; Jeong et al., 2015; Plaut, 1954a; Plaut, 1954b; Plaut and Broberg, 1956; Schlüpen et al., 2003). For the first time, the results presented here gave an insight into the metabolism of the fungus under industrially relevant conditions (Sahm et al., 2013), which uses a complex medium with vegetable oil as main carbon source, supplemented with yeast extract and glycine.

A detailed analysis of carbon fluxes during growth and riboflavin production in a very complex cultivation set-up using *A. gossypii* was presented in this work. The ubiquitous industrial medium ingredient yeast extract was ^{13}C-labeled for the first time and thus, its impact on the metabolism studied in greater detail. Yeast extract could be identified as the main carbon source during growth, while rapeseed oil was the main carbon source during riboflavin production. However, the ^{13}C labeling data and resulting carbon fluxes demonstrated that yeast extract also contributed a significant amount of carbon to the product. Other industrial processes also rely heavily on yeast extract as complex nitrogen source (Papagianni, 2004; Zhang et al., 2003). The commercial production of L-lysine with bacteria uses yeast extract as medium ingredient (Kojima et al., 2000) as well as the antibiotics fermentation with filamentous fungi (Papagianni, 2004; Posch et al., 2012). *Bacillus thuringiensis* is industrially employed as insecticide producer, a process which depends on complex compounds such as corn steep liquor or soybean meal. Since those ingredients might not be readily available at every production site, yeast extract was tested as alternative complex nutrient (Saksinchai et al., 2001). These studies highlight the importance of complex media for industrial fermentation processes and the need for a more sophisticated experimental set-up in order to study and improve such systems in great detail. The ^{13}C labeling approach presented here, provides an excellent and promising starting point for the improvement of other microbial processes under complex and commercial cultivation conditions.

Regarding riboflavin biosynthesis with *A. gossypii* there are still several issues that need to be investigated further. As presented in this work, intracellular availability of riboflavin precursors is crucial. Along with that intracellular localization of metabolites clearly has an effect on productivity. Regarding *A. gossypii*, little is known about the exact transport mechanisms for amino acids or formate, whether transport across the plasma membrane or transport in and out of the vacuole are concerned. *A. gossypii* and *S. cerevisiae* were reported to store large amounts of glycine in their vacuole (Förster et al., 1998; Messenguy et al., 1980), thus making it inaccessible for riboflavin production. Here, fine-tuned transport systems might overcome this limitation. Disruption of vacuolar transport of riboflavin successfully increased product titers (Förster et al., 1999), but the transport mechanism across the plasma membrane still needs to be elucidated. Recently, new bacterial riboflavin transporters have been identified and characterized (Gutiérrez-Preciado et al., 2015), however, the process and regulation of riboflavin secretion into the culture medium still needs to be further investigated.

The results presented here, identified glutamate as a key player in the metabolism, however, it only slightly contributed to biomass and riboflavin. With regard to the four nitrogen atoms present in riboflavin, it would be very interesting to resolve the underlying metabolism. Urea, which was present in the medium in large amounts, certainly also feeds the intracellular nitrogen pool. In light of environmental awareness and hazards, optimization of producing strains to use second and third generation raw materials might be of great research interest. As example, *A. gossypii* is able to utilize waste activated bleaching earth, a waste disposal of vegetable oil manufacturers. Riboflavin titers of 1 to 2 g L^{-1} were reported for *A. gossypii* using these solid waste substrates compared to rapeseed or palm oil (Ming et al., 2003; Park et al., 2004; Park and Ming, 2004).

6 APPENDIX

6.1 Abbreviations

(10-)CHO-THF	10-Formyltetrahydrofolate
(5,10-)CH$_2$-THF	5,10-Methylenetetrahydrofolate
3PG	3-Phosphoglycerate
AAT	Gene encoding aspartate transaminase
Ac	Acetate
ACES	*N*-(2-Acetamido)-2-aminoethanesulfonic acid
Acetal	Acetaldehyde
Acetyl-CoA/AcCoA	Acetyl coenzyme A
acs	Gene encoding acetyl-CoA-synthetase
ADE1	Gene encoding phosphoribosylaminoimidazole-succinocarboxamide synthase
ADE12	Gene encoding adenylosuccinate synthase
ADE4	Gene encoding PRPP amidotransferase
ADY2	Gene encoding potential acetate transporter
AGX1	Gene encoding alanine-glyoxylate aminotransferase
Ala	Alanine
ALT1/2	Gene encoding alanine transaminase 1/2
AMP	Adenosine monophosphate
Arg	Arginine
ArP	5-Amino-6-ribitylamino-2,4(1H,3H)-pyrimidinedione
ArPP	5-amino-6-ribitylamino-2,4(1H,3H)-pyrimidinedione 5-phosphate
Asp	Aspartic acid
ATCC	American type culture collection
ATP	Adenosine triphosphate
BAS1	Gene encoding *Myb*-related transcription factor
BSTFA	*N,O-Bis*-trimethylsilyl-trifluoroacetamide
cAMP	Cyclic adenosine monophosphate
ccpn	Gene encoding transcriptional regulator of *gapB* and *pckA*
CDW	Cell dry weight
Cit	Citric acid
CTP	Cytidine triphosphate
cyd	Gene encoding cytochrome *bd* oxidase
DAK1/2	Genes encoding dihydroxyacetone kinase 1/2
DArPP	2,5-Diamino-6-ribitylamino-pyrimidinone
DARPP	2,5-Diamino-6-ribosylamino-4(3H)-pyrimidinone
dATP	Deoxyadenosine triphosphate
dCTP	Deoxycytidine triphosphate
dGTP	Deoxyguanosine triphosphate
DHAP	Dihydroxyacetone phosphate

DHBP	3,4-Dihydroxy-2-butanone 4-phosphate
DMF	Dimethylformamide
DRL	6,7-Dimethyl-8-ribityllumazine
dTTP	Deoxythymidine triphosphate
E4P	Erythrose 4-phosphate
eda	Gene encoding multifunctional 2-keto-3-deoxygluconate 6-phosphate aldolase and 2-keto-4-hydroxyglutarate aldolase and oxaloacetate decarboxylase
edd	Gene encoding phosphogluconate dehydratase
EMU	Elementary mode units
F6P	Fructose 6-phosphae
FA	Fatty acid
FAD	Flavin adenine dinucleotide
fbp	Gene encoding fructose-1,6-bisphosphatase
FID	Free induction decay
FMN	Flavin mononucleotide
For	Formic acid
G3P	Glyceraldehyde 3-phosphate
G6P	Glucose 6-phosphate
GAP1	Gene encoding general amino acid permease
gapB	Gene encoding glyceraldehyde-3-phosphate dehydrogenase
GAR	Glycineamide ribonucleotide
GC/MS	Gas chromatography/mass spectrometry
GCS	Glycine cleavage system
gdh	Gene encoding glucose dehydrogenase
GDH1/3	Gene encoding NADP-dependent glutamate dehydrogenase 1/3
GDH2	Gene encoding NAD-dependent glutamate dehydrogenase
Glc	Glucose
Glu	Glutamic acid
Gly	Glycine
GLY1	Gene encoding threonine aldolase
Glyc	Glycerol
Glyox	Glyoxylate
GMP	Guanosine monophosphate
gnd	Gene encoding 6-phosphogluconate dehydrogenase
GTP	Guanosine triphosphate
Gua	Guanine
guaA	Gene encoding GMP synthase
guaB	Gene encoding IMP dehydrogenase
GUT1	Gene encoding glycerol kinase
GUT2	Gene encoding FAD-dependent glycerol 3-phosphate dehydrogenase
HCl	Hydrogen chloride
His	Histidine
HPLC	High performance liquid chromatography
HXT	Genes encoding hexose transporters
Icit	Isocitric acid

ICL	Isocitrate lyase
IDH	Isocitrate dehydrogenase
Ile	Isoleucine
IMP	Inosine monophosphate
IMPDH	Gene encoding IMP dehydrogenase
JEN1	Gene encoding an acetate permease
KEGG	Kyoto Encyclopedia of Genes and Genomes
LC/MS	Liquid chromatography/mass spectrometry
Leu	Leucine
Lys	Lysine
MaE	Malic enzyme
Mal	Malate
MBDSTFA	*N*-Methyl-*N*-tert-butyl-dimethylsilyl-trifluoroacetamide
MFA	Metabolic flux analysis
MFS	Major facilitator superfamily
MID	Mass isotopomer distribution
MLS1	Gene encoding malate synthase
MOPS	3-(*N*-Morpholino)propanesulfonic acid
NAD/NADH	Nicotinamide adenine dinucleotide ox/red
NADP/NADPH	Nicotinamide adenine dinucleotide phosphate ox/red
NaOH	Sodium hydroxide
NMR	Nuclear magnetic resonance
OAA	Oxaloacetic acid
OMP	Orotidine monophosphate
P5P	Pentose 5-phosphate
PEP	Phosphoenolpyruvate
pgi	Gene encoding glucose-6-phosphate isomerase
pgl	Gene encoding 6-phosphogluconolactonase
Phe	Phenylalanine
PP pathway	Pentose phosphate pathway
PRA	5-Phosphoribosylamine
Pro	Proline
PRPP	Phosphoribosylpyrophosphate
PRS	Gene encoding PRPP synthetase
PTS	Phosphotransferase system
purA	Gene encoding adenylosuccinate synthetase
purD	Gene encoding phosphoribosylglycinamide synthetase
purF	Gene encoding glutamine phosphoribosylpyrophosphate amidotransferase
purH	Gene encoding phosphoribosylaminoimidazole carboxy formylformyltransferase/IMP cyclohydrolase
purM	Gene encoding phosphoribosylaminoimidazole synthetase
purN	Gene encoding phosphoribosylglycinamide formyltransferase
purR	Gene encoding transcriptional repressor of *pur* operon
Pyr	Pyruvate
R5P	Ribose 5-phosphate

RF	Riboflavin
RIB1	Gene encoding GTP cyclohydrolase II
RIB2	Gene encoding DArPP deaminase
RIB3	Gene encoding DHBP synthase
RIB4	Gene encoding lumazine synthase
RIB5	Gene encoding riboflavin synthase
RIB7	Gene encoding DARPP reductase
ribA	Gene encoding GTP cyclohydrolase II and DHBP synthase
ribC	*Trans*-acting regulator of the *rib* operon
ribF	Gene encoding riboflavin kinase
ribG	Gene encoding DARPP reductase and DArPP deaminase
ribGBAHT	*rib* operon
ribR	Regulatory gene of *rib* operon
rpm	Revolutions per minute
RQ	Respiration quotient
Ru5P	Ribulose 5-phosphate
S7P	Sedoheptulose 7-phosphate
SCO	Single cell oil
SEP	3-Phosphoserine
Ser	Serine
SFL	Summed fractional labeling
SHM	Gene encoding serine hydroxymethyltransferase
SHMT	Serine hydroxymethyltransferase
sigH	Gene encoding sigma factor H
SIM	Selective ion monitoring
SP	Sporulation
Sug	Sugar
TBDMS	*tert*-Butyl-dimethylsilyl
TCA cycle	Tricarboxylic acid cycle
THF	Tetrahydrofolate
Thr	Threonine
TMS	Trimethylsilyl
TSP-d4	2-(Trimethylsilyl)propionic-2,2,3,3-d4 acid
Tyr	Tyrosine
UK	United Kingdom
UMP	Uridine monophosphate
URA3	Gene encoding orotidine-5-phosphate decarboxylase
US(A)	United States (of America)
UTP	Uridine triphosphate
UV	Ultraviolet
Val	Valine
VMA1	Gene encoding vacuolar ATPase subunit A
X	Biomass
XMP	Xanthosine monophosphate
Xu5P	Xylulose 5-phosphate

YAP1	Gene encoding transcriptional regulator
YE	Yeast extract
ywlF	Gene encoding ribose-5-phosphate isomerase

6.2   Symbols

$\%^{13}C$	Positional ^{13}C enrichment of a carbon atom	[%]
[RF]	Riboflavin concentration	[g L^{-1}]
µ	Specific growth rate	[h^{-1}]
c	Concentration	[g L^{-1}] or [mol L^{-1}]
m	Mass	[g]
m/z	Mass-to-charge ratio	[-]
P	Purity of a ^{13}C tracer	[%]
q_P	Specific product formation rate	[µmol g^{-1} h^{-1}]
Q_P	Volumetric product formation rate	[µmol L^{-1} h^{-1}]
q_S	Specific substrate uptake rate	[mmol g^{-1} h^{-1}]
T	Temperature	[°C]
t	Time	[h]
v	Flux	[%] or [mmol g^{-1} h^{-1}] or [Carbon atom]
$Y_{P/S}$	Product yield	[mol mol^{-1}]
$Y_{X/S}$	Biomass yield	[g mol^{-1}]
ζ	Flux reversibility	[-]
λ	Wavelenght	[nm]
Ψ_{TR}	Contribution of a ^{13}C tracer to a target molecule	[%]

6.3   Isotope experiments conducted in this study

Table 15: Overview of ^{13}C tracer studies performed in this work. Time point refers to the tracer addition. All cultivations were carried out in complex medium and glucose (No. 1, 2), glycerol:acetate (No. 3-5) or rapeseed oil (No. 6-16) as carbon source.

No.	Tracer	Time point [h]	Strain	Reference
1	[$^{13}C_6$] Glucose	0	WT	Chapter 4.2
2	[$^{13}C_6$] Glucose	0	B2	Chapter 4.2
3	[$^{13}C_3$] Glycerol	0	WT	Chapter 4.3
4	[$^{13}C_2$] Sodium acetate	0	B2	Chapter 4.3
5	[$^{13}C_3$] Glycerol	0	WT	Chapter 4.3
6	[$^{13}C_2$] Sodium acetate	0	B2	Chapter 4.3
7	[$^{13}C_2$] Glycine	0	B2	Chapters 4.4, 4.5
8	[$^{13}C_3$] Serine	0	B2	Chapters 4.4, 4.5
9	[$^{13}C_3$] Serine	48	B2	Chapter 4.5
10	[3-^{13}C] Serine	0	B2	Chapter 4.5
11	[3-^{13}C] Serine	48	B2	Chapter 4.5
12	[^{13}C] Sodium formate	0	B2	Chapters 4.4, 4.5
13	[^{13}C] Sodium formate	48	B2	Chapter 4.5
14	[U^{13}C] Yeast extract	0 – 32[a]	B2	Chapter 4.4
15	[U^{13}C] Yeast extract	32 – 144[b]	B2	Chapters 4.5
16	[$^{13}C_5$] Glutamic acid	0	B2	Chapters 4.4, 4.5
17	[U^{13}C] Yeast extract, [$^{13}C_2$] glycine, [$^{13}C_5$] glutamic acid, [^{13}C] sodium formate	0	B2	Chapters 4.4, 4.5

[a] Medium was exchanged with naturally labeled medium after 32 h
[b] Naturally labeled medium was exchanged with [U^{13}C] yeast extract containing medium after 32 h

6.4   Data from GC/MS analyses

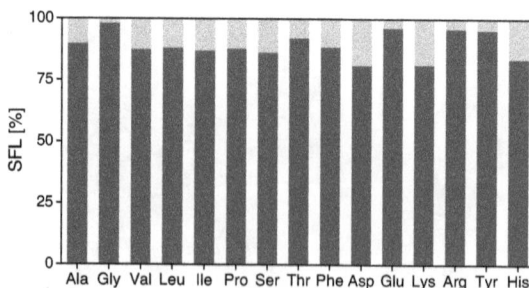

Figure 39: Summed fractional labeling (SFL) derived from GC/MS measurements of the culture supernatant of *A. gossypii* B2 cells grown on complex medium with [U^{13}C] yeast extract, [$^{13}C_2$] glycine, [^{13}C] formate, and [$^{13}C_5$] glutamate in a single experiment. Labeling data derived from that cultivation are depicted in dark grey while the light grey bars indicate the resulting contribution of non-labeled pre-culture medium to the supernatant. Data were obtained from three individual replicates and denote mean values with a mean standard deviation of 5 %.

Table 16: Relative mass isotopomer fractions of amino acids from hydrolyzed cell protein of *A. gossypii* WT and B2 grown on naturally labeled complex medium and 99 % [$^{13}C_6$] glucose, [$^{13}C_2$] acetate, or [$^{13}C_3$] glycerol. Data denote corrected labeling patterns. Data were corrected for fraction of unlabeled biomass in the inoculum as well as occurrence of natural isotopes. The mass isotopomer M+0 represents the relative amount of non-labeled, M+1 the amount of singly-labeled mass isotopomer fraction and so on. Data were obtained from three individual replicates.

Analyte		WT [$^{13}C_6$] Glc	WT [$^{13}C_2$] Ac	WT [$^{13}C_3$] Glyc	B2 [$^{13}C_6$] Glc	B2 [$^{13}C_2$] Ac	B2 [$^{13}C_3$] Glyc
Ala_260	M+0	0.40 ± 0.02	0.77 ± 0.05	0.72 ± 0.04	0.36 ± 0.02	0.75 ± 0.07	0.84 ± 0.06
	M+1	0.02 ± 0.00	0.05 ± 0.00	0.04 ± 0.00	0.01 ± 0.00	0.05 ± 0.03	0.04 ± 0.02
	M+2	0.02 ± 0.00	0.06 ± 0.00	0.01 ± 0.00	0.02 ± 0.00	0.06 ± 0.03	0.01 ± 0.00
	M+3	0.56 ± 0.03	0.12 ± 0.01	0.23 ± 0.01	0.60 ± 0.04	0.14 ± 0.10	0.11 ± 0.01
	SFL [%]	57.81 ± 2.71	17.31 ± 1.21	25.26 ± 1.41	61.92 ± 3.44	19.42 ± 1.90	12.77 ± 0.91
	SFL$_{corr}$ [%]	**57.31 ± 2.69**	**16.40 ± 1.04**	**24.44 ± 1.36**	**61.46 ± 3.41**	**18.53 ± 1.81**	**11.82 ± 0.84**
Gly_246	M+0	0.97 ± 0.00	0.97 ± 0.00	0.98 ± 0.00	0.97 ± 0.00	0.97 ± 0.00	0.98 ± 0.00
	M+1	0.03 ± 0.00	0.02 ± 0.00	0.02 ± 0.00	0.02 ± 0.00	0.02 ± 0.00	0.02 ± 0.00
	M+2	0.01 ± 0.00	0.01 ± 0.00	0.00 ± 0.00	0.00 ± 0.00	0.00 ± 0.00	0.00 ± 0.00
	SFL [%]	2.15 ± 0.10	1.69 ± 0.07	1.23 ± 0.06	1.46 ± 0.08	1.48 ± 0.08	1.20 ± 0.05
	SFL$_{corr}$ [%]	**1.09 ± 0.05**	**0.62 ± 0.03**	**0.16 ± 0.01**	**0.39 ± 0.02**	**0.42 ± 0.02**	**0.13 ± 0.01**
Val_288	M+0	0.91 ± 0.02	0.94 ± 0.00	0.94 ± 0.00	0.94 ± 0.00	0.94 ± 0.00	0.95 ± 0.01
	M+1	0.05 ± 0.00	0.05 ± 0.00	0.05 ± 0.00	0.05 ± 0.00	0.05 ± 0.00	0.05 ± 0.00
	M+2	0.00 ± 0.00	0.00 ± 0.00	0.00 ± 0.00	0.00 ± 0.00	0.00 ± 0.00	0.00 ± 0.00
	M+3	0.00 ± 0.00	0.00 ± 0.00	0.00 ± 0.00	0.00 ± 0.00	0.00 ± 0.00	0.00 ± 0.00
	M+4	0.00 ± 0.00	0.00 ± 0.00	0.00 ± 0.00	0.00 ± 0.00	0.00 ± 0.00	0.00 ± 0.00
	M+5	0.03 ± 0.00	0.00 ± 0.00	0.00 ± 0.00	0.01 ± 0.00	0.00 ± 0.00	0.00 ± 0.00
	SFL [%]	4.52 ± 0.41	1.39 ± 0.10	1.41 ± 0.13	1.79 ± 0.13	1.44 ± 0.14	1.30 ± 0.12
	SFL$_{corr}$ [%]	**3.49 ± 0.32**	**0.32 ± 0.02**	**0.34 ± 0.02**	**0.72 ± 0.05**	**0.38 ± 0.04**	**0.23 ± 0.02**
Leu_274	M+0	0.94 ± 0.00	0.94 ± 0.00	0.95 ± 0.00	0.95 ± 0.00	0.95 ± 0.00	0.95 ± 0.00
	M+1	0.06 ± 0.00	0.05 ± 0.00	0.05 ± 0.00	0.05 ± 0.00	0.05 ± 0.00	0.05 ± 0.00
	M+2	0.00 ± 0.00	0.00 ± 0.00	0.00 ± 0.00	0.00 ± 0.00	0.00 ± 0.00	0.00 ± 0.00
	M+3	0.00 ± 0.00	0.00 ± 0.00	0.00 ± 0.00	0.00 ± 0.00	0.00 ± 0.00	0.00 ± 0.00
	M+4	0.00 ± 0.00	0.00 ± 0.00	0.00 ± 0.00	0.00 ± 0.00	0.00 ± 0.00	0.00 ± 0.00
	M+5	0.00 ± 0.00	0.00 ± 0.00	0.00 ± 0.00	0.00 ± 0.00	0.00 ± 0.00	0.00 ± 0.00
	SFL [%]	1.12 ± 0.05	1.13 ± 0.06	1.08 ± 0.00	1.09 ± 0.01	1.12 ± 0.04	1.10 ± 0.03
	SFL$_{corr}$ [%]	**0.13 ± 0.01**	**0.08 ± 0.00**	**0.08 ± 0.00**	**0.07 ± 0.00**	**0.09 ± 0.00**	**0.08 ± 0.00**
Ile_274	M+0	0.92 ± 0.01	0.93 ± 0.00	0.94 ± 0.00	0.93 ± 0.00	0.93 ± 0.00	0.94 ± 0.00
	M+1	0.07 ± 0.00	0.06 ± 0.00	0.06 ± 0.00	0.07 ± 0.00	0.06 ± 0.00	0.06 ± 0.00
	M+2	0.01 ± 0.00	0.00 ± 0.00	0.00 ± 0.00	0.01 ± 0.00	0.00 ± 0.00	0.00 ± 0.00
	M+3	0.00 ± 0.00	0.00 ± 0.00	0.00 ± 0.00	0.00 ± 0.00	0.00 ± 0.00	0.00 ± 0.00
	M+4	0.00 ± 0.00	0.00 ± 0.00	0.00 ± 0.00	0.00 ± 0.00	0.00 ± 0.00	0.00 ± 0.00
	M+5	0.00 ± 0.00	0.00 ± 0.00	0.00 ± 0.00	0.00 ± 0.00	0.00 ± 0.00	0.00 ± 0.00
	SFL [%]	1.75 ± 0.12	1.47 ± 0.10	1.51 ± 0.06	1.64 ± 0.16	1.53 ± 0.14	1.39 ± 0.11
	SFL$_{corr}$ [%]	**0.24 ± 0.02**	**0.14 ± 0.01**	**0.24 ± 0.01**	**0.19 ± 0.02**	**0.27 ± 0.02**	**0.33 ± 0.03**

Table 16: Relative mass isotopomer fractions of amino acids from hydrolyzed cell protein of *A. gossypii* WT and B2 grown on naturally labeled complex medium and 99 % [$^{13}C_6$] glucose, [$^{13}C_2$] acetate, or [$^{13}C_3$] glycerol **(continued)**.

Analyte		WT			B2		
		[$^{13}C_6$] Glc	[$^{13}C_2$] Ac	[$^{13}C_3$] Glyc	[$^{13}C_6$] Glc	[$^{13}C_2$] Ac	[$^{13}C_3$] Glyc
Pro_258	M+0	0.90 ± 0.01	n.d.	n.d.	0.91 ± 0.01	n.d.	n.d.
	M+1	0.04 ± 0.00	n.d.	n.d.	0.04 ± 0.00	n.d.	n.d.
	M+2	0.01 ± 0.00	n.d.	n.d.	0.01 ± 0.00	n.d.	n.d.
	M+3	0.01 ± 0.00	n.d.	n.d.	0.01 ± 0.00	n.d.	n.d.
	M+4	0.04 ± 0.00	n.d.	n.d.	0.03 ± 0.00	n.d.	n.d.
	SFL [%]	5.81 ± 0.21	n.d.	n.d.	5.19 ± 0.30	n.d.	n.d.
	SFL$_{corr}$ [%]	**4.79 ± 0.17**	**n.d.**	**n.d.**	**4.16 ± 0.24**	**n.d.**	**n.d.**
Ser_390	M+0	0.90 ± 0.01	0.94 ± 0.00	0.97 ± 0.00	0.96 ± 0.00	0.94 ± 0.00	0.97 ± 0.00
	M+1	0.05 ± 0.00	0.06 ± 0.00	0.03 ± 0.00	0.03 ± 0.00	0.05 ± 0.00	0.03 ± 0.00
	M+2	0.01 ± 0.00	0.00 ± 0.00	0.00 ± 0.00	0.00 ± 0.00	0.00 ± 0.00	0.00 ± 0.00
	M+3	0.05 ± 0.00	0.00 ± 0.00	0.00 ± 0.00	0.00 ± 0.00	0.00 ± 0.00	0.00 ± 0.00
	SFL [%]	6.90 ± 0.51	2.34 ± 0.10	1.27 ± 0.08	1.36 ± 0.09	2.02 ± 0.11	1.25 ± 0.08
	SFL$_{corr}$ [%]	**5.89 ± 0.44**	**1.29 ± 0.06**	**0.20 ± 0.01**	**0.30 ± 0.02**	**0.96 ± 0.05**	**0.18 ± 0.01**
Thr_404	M+0	0.90 ± 0.02	0.96 ± 0.00	0.96 ± 0.00	0.94 ± 0.01	0.96 ± 0.00	0.96 ± 0.00
	M+1	0.04 ± 0.00	0.04 ± 0.00	0.04 ± 0.00	0.04 ± 0.00	0.04 ± 0.00	0.04 ± 0.00
	M+2	0.05 ± 0.00	0.00 ± 0.00	0.00 ± 0.00	0.02 ± 0.00	0.00 ± 0.00	0.00 ± 0.00
	M+3	0.00 ± 0.00	0.00 ± 0.00	0.00 ± 0.00	0.00 ± 0.00	0.00 ± 0.00	0.00 ± 0.00
	M+4	0.01 ± 0.00	0.00 ± 0.00	0.00 ± 0.00	0.00 ± 0.00	0.00 ± 0.00	0.00 ± 0.00
	SFL [%]	4.26 ± 0.22	1.08 ± 0.01	0.92 ± 0.10	2.43 ± 0.10	1.50 ± 0.11	1.11 ± 0.05
	SFL$_{corr}$ [%]	**3.23 ± 0.17**	**0.02 ± 0.00**	**-0.16 ± 0.02**	**1.38 ± 0.06**	**0.43 ± 0.03**	**0.04 ± 0.00**
Phe_336	M+0	0.91 ± 0.00	0.91 ± 0.00	0.91 ± 0.00	0.91 ± 0.00	0.91 ± 0.00	0.91 ± 0.00
	M+1	0.09 ± 0.00	0.09 ± 0.00	0.09 ± 0.00	0.09 ± 0.00	0.09 ± 0.00	0.09 ± 0.00
	M+2	0.00 ± 0.00	0.00 ± 0.00	0.00 ± 0.00	0.00 ± 0.00	0.00 ± 0.00	0.00 ± 0.00
	M+3	0.00 ± 0.00	0.00 ± 0.00	0.00 ± 0.00	0.00 ± 0.00	0.00 ± 0.00	0.00 ± 0.00
	M+4	0.00 ± 0.00	0.00 ± 0.00	0.00 ± 0.00	0.00 ± 0.00	0.00 ± 0.00	0.00 ± 0.00
	M+5	0.00 ± 0.00	0.00 ± 0.00	0.00 ± 0.00	0.00 ± 0.00	0.00 ± 0.00	0.00 ± 0.00
	M+6	0.00 ± 0.00	0.00 ± 0.00	0.00 ± 0.00	0.00 ± 0.00	0.00 ± 0.00	0.00 ± 0.00
	M+7	0.00 ± 0.00	0.00 ± 0.00	0.00 ± 0.00	0.00 ± 0.00	0.00 ± 0.00	0.00 ± 0.00
	M+8	0.00 ± 0.00	0.00 ± 0.00	0.00 ± 0.00	0.00 ± 0.00	0.00 ± 0.00	0.00 ± 0.00
	M+9	0.00 ± 0.00	0.00 ± 0.00	0.00 ± 0.00	0.00 ± 0.00	0.00 ± 0.00	0.00 ± 0.00
	SFL [%]	1.09 ± 0.01	1.07 ± 0.00	1.23 ± 0.10	1.09 ± 0.03	1.10 ± 0.03	1.12 ± 0.06
	SFL$_{corr}$ [%]	**0.02 ± 0.00**	**0.00 ± 0.00**	**0.16 ± 0.01**	**0.02 ± 0.00**	**0.03 ± 0.00**	**0.05 ± 0.00**
Asp_418	M+0	0.45 ± 0.04	0.47 ± 0.03	0.86 ± 0.04	0.54 ± 0.03	0.50 ± 0.04	0.93 ± 0.01
	M+1	0.03 ± 0.00	0.04 ± 0.00	0.08 ± 0.00	0.03 ± 0.00	0.05 ± 0.00	0.05 ± 0.00
	M+2	0.04 ± 0.00	0.09 ± 0.00	0.04 ± 0.00	0.03 ± 0.00	0.08 ± 0.00	0.01 ± 0.00
	M+3	0.10 ± 0.00	0.12 ± 0.01	0.02 ± 0.00	0.08 ± 0.00	0.11 ± 0.00	0.01 ± 0.00
	M+4	0.39 ± 0.02	0.27 ± 0.02	0.00 ± 0.00	0.33 ± 0.03	0.26 ± 0.01	0.00 ± 0.00
	SFL [%]	48.96 ± 3.45	41.91 ± 2.96	5.62 ± 2.33	40.59 ± 2.26	39.75 ± 3.18	2.53 ± 0.03
	SFL$_{corr}$ [%]	**48.37 ± 3.41**	**41.26 ± 2.91**	**4.59 ± 1.90**	**39.92 ± 2.22**	**39.07 ± 3.73**	**1.47 ± 0.02**

Table 16: Relative mass isotopomer fractions of amino acids from hydrolyzed cell protein of *A. gossypii* WT and B2 grown on naturally labeled complex medium and 99 % [$^{13}C_6$] glucose, [$^{13}C_2$] acetate, or [$^{13}C_3$] glycerol **(continued)**.

Analyte		WT [$^{13}C_6$] Glc	WT [$^{13}C_2$] Ac	WT [$^{13}C_3$] Glyc	B2 [$^{13}C_6$] Glc	B2 [$^{13}C_2$] Ac	B2 [$^{13}C_3$] Glyc
Glu_432	M+0	0.14 ± 0.01	0.20 ± 0.01	0.80 ± 0.01	0.20 ± 0.01	0.17 ± 0.01	0.90 ± 0.01
	M+1	0.01 ± 0.00	0.02 ± 0.00	0.10 ± 0.00	0.01 ± 0.00	0.02 ± 0.00	0.07 ± 0.00
	M+2	0.08 ± 0.00	0.11 ± 0.00	0.07 ± 0.00	0.08 ± 0.00	0.11 ± 0.00	0.03 ± 0.00
	M+3	0.08 ± 0.00	0.10 ± 0.00	0.01 ± 0.00	0.07 ± 0.00	0.11 ± 0.00	0.00 ± 0.00
	M+4	0.15 ± 0.00	0.18 ± 0.00	0.00 ± 0.00	0.15 ± 0.00	0.18 ± 0.00	0.00 ± 0.00
	M+5	0.54 ± 0.04	0.39 ± 0.03	0.00 ± 0.00	0.49 ± 0.02	0.41 ± 0.04	0.00 ± 0.00
	SFL [%]	74.43 ± 6.45	64.08 ± 6.30	6.56 ± 2.11	68.29 ± 3.85	66.63 ± 6.70	2.95 ± 0.05
	SFL$_{corr}$ [%]	**74.10 ± 6.42**	**63.65 ± 6.24**	**5.55 ± 1.77**	**67.90 ± 3.83**	**66.22 ± 6.66**	**1.90 ± 0.03**
Lys_431	M+0	0.94 ± 0.00	0.94 ± 0.00	0.95 ± 0.01	0.94 ± 0.00	0.94 ± 0.00	0.94 ± 0.00
	M+1	0.06 ± 0.00	0.06 ± 0.00	0.05 ± 0.00	0.06 ± 0.00	0.06 ± 0.00	0.06 ± 0.00
	M+2	0.00 ± 0.00	0.00 ± 0.00	0.00 ± 0.00	0.00 ± 0.00	0.00 ± 0.00	0.00 ± 0.00
	M+3	0.00 ± 0.00	0.00 ± 0.00	0.00 ± 0.00	0.00 ± 0.00	0.00 ± 0.00	0.00 ± 0.00
	M+4	0.00 ± 0.00	0.00 ± 0.00	0.00 ± 0.00	0.00 ± 0.00	0.00 ± 0.00	0.00 ± 0.00
	M+5	0.00 ± 0.00	0.00 ± 0.00	0.00 ± 0.00	0.00 ± 0.00	0.00 ± 0.00	0.00 ± 0.00
	M+6	0.00 ± 0.00	0.00 ± 0.00	0.00 ± 0.00	0.00 ± 0.00	0.00 ± 0.00	0.00 ± 0.00
	SFL [%]	1.06 ± 0.01	1.01 ± 0.05	0.89 ± 0.10	0.96 ± 0.08	0.98 ± 0.08	0.95 ± 0.10
	SFL$_{corr}$ [%]	**-0.01 ± 0.00**	**-0.06 ± 0.00**	**-0.18 ± 0.02**	**-0.11 ± 0.01**	**-0.09 ± 0.01**	**-0.12 ± 0.01**
Arg_442	M+0	0.91 ± 0.02	0.94 ± 0.00	0.94 ± 0.00	0.94 ± 0.00	0.94 ± 0.00	0.94 ± 0.00
	M+1	0.06 ± 0.00	0.05 ± 0.00	0.06 ± 0.00	0.05 ± 0.00	0.05 ± 0.00	0.06 ± 0.00
	M+2	0.00 ± 0.00	0.00 ± 0.00	0.00 ± 0.00	0.00 ± 0.00	0.00 ± 0.00	0.00 ± 0.00
	M+3	0.00 ± 0.00	0.00 ± 0.00	0.00 ± 0.00	0.00 ± 0.00	0.00 ± 0.00	0.00 ± 0.00
	M+4	0.00 ± 0.00	0.00 ± 0.00	0.00 ± 0.00	0.00 ± 0.00	0.00 ± 0.00	0.00 ± 0.00
	M+5	0.01 ± 0.00	0.00 ± 0.00	0.00 ± 0.00	0.00 ± 0.00	0.00 ± 0.00	0.00 ± 0.00
	M+6	0.01 ± 0.00	0.00 ± 0.00	0.00 ± 0.00	0.00 ± 0.00	0.00 ± 0.00	0.00 ± 0.00
	SFL [%]	3.19 ± 0.20	1.01 ± 0.06	1.15 ± 0.07	1.21 ± 0.10	1.16 ± 0.08	1.01 ± 0.06
	SFL$_{corr}$ [%]	**2.14 ± 0.13**	**-0.06 ± 0.00**	**0.08 ± 0.00**	**0.14 ± 0.01**	**0.09 ± 0.01**	**-0.06 ± 0.00**
Tyr_466	M+0	0.63 ± 0.04	0.79 ± 0.05	0.74 ± 0.03	0.73 ± 0.04	0.84 ± 0.01	0.83 ± 0.02
	M+1	0.05 ± 0.00	0.09 ± 0.00	0.07 ± 0.00	0.06 ± 0.00	0.08 ± 0.00	0.08 ± 0.00
	M+2	0.00 ± 0.00	0.03 ± 0.00	0.01 ± 0.00	0.00 ± 0.00	0.02 ± 0.00	0.01 ± 0.00
	M+3	0.00 ± 0.00	0.03 ± 0.00	0.02 ± 0.00	0.00 ± 0.00	0.02 ± 0.00	0.02 ± 0.00
	M+4	0.00 ± 0.00	0.02 ± 0.00	0.02 ± 0.00	0.00 ± 0.00	0.01 ± 0.00	0.01 ± 0.00
	M+5	0.00 ± 0.00	0.02 ± 0.00	0.03 ± 0.00	0.00 ± 0.00	0.01 ± 0.00	0.01 ± 0.00
	M+6	0.00 ± 0.00	0.01 ± 0.00	0.03 ± 0.00	0.00 ± 0.00	0.01 ± 0.00	0.01 ± 0.00
	M+7	0.01 ± 0.00	0.01 ± 0.00	0.02 ± 0.00	0.00 ± 0.00	0.00 ± 0.00	0.01 ± 0.00
	M+8	0.03 ± 0.00	0.01 ± 0.00	0.02 ± 0.00	0.02 ± 0.00	0.00 ± 0.00	0.00 ± 0.00
	M+9	0.27 ± 0.02	0.00 ± 0.00	0.04 ± 0.00	0.18 ± 0.01	0.00 ± 0.00	0.01 ± 0.00
	SFL [%]	31.35 ± 1.99	6.54 ± 0.41	13.93 ± 0.56	20.90 ± 1.15	4.65 ± 0.06	5.56 ± 0.13
	SFL$_{corr}$ [%]	**30.59 ± 1.94**	**5.53 ± 0.35**	**12.99 ± 0.52**	**20.03 ± 1.10**	**3.61 ± 0.05**	**4.54 ± 0.11**

Table 16: Relative mass isotopomer fractions of amino acids from hydrolyzed cell protein of *A. gossypii* WT and B2 grown on naturally labeled complex medium and 99 % [$^{13}C_6$] glucose, [$^{13}C_2$] acetate, or [$^{13}C_3$] glycerol **(continued)**.

Analyte		WT			B2		
		[$^{13}C_6$] Glc	[$^{13}C_2$] Ac	[$^{13}C_3$] Glyc	[$^{13}C_6$] Glc	[$^{13}C_2$] Ac	[$^{13}C_3$] Glyc
His_440	M+0	0.90 ± 0.01	0.92 ± 0.00	n.d.	0.74 ± 0.03	0.92 ± 0.00	0.93 ± 0.00
	M+1	0.07 ± 0.00	0.07 ± 0.00	n.d.	0.15 ± 0.01	0.07 ± 0.00	0.05 ± 0.00
	M+2	0.01 ± 0.00	0.01 ± 0.00	n.d.	0.05 ± 0.00	0.01 ± 0.00	0.01 ± 0.00
	M+3	0.01 ± 0.00	0.00 ± 0.00	n.d.	0.03 ± 0.00	0.01 ± 0.00	0.00 ± 0.00
	M+4	0.00 ± 0.00	0.00 ± 0.00	n.d.	0.02 ± 0.00	0.00 ± 0.00	0.00 ± 0.00
	M+5	0.00 ± 0.00	0.00 ± 0.00	n.d.	0.00 ± 0.00	0.00 ± 0.00	0.00 ± 0.00
	M+6	0.00 ± 0.00	0.00 ± 0.00	n.d.	0.01 ± 0.00	0.00 ± 0.00	0.00 ± 0.00
	SFL [%]	2.42 ± 0.04	1.53 ± 0.01	n.d.	7.61 ± 0.35	1.87 ± 0.02	1.61 ± 0.02
	SFL$_{corr}$ [%]	**1.36 ± 0.02**	**0.47 ± 0.00**	**n.d.**	**6.60 ± 0.30**	**0.81 ± 0.01**	**0.54 ± 0.01**

Table 17: Relative mass isotopomer fractions of amino acids from hydrolyzed cell protein of *A. gossypii* B2 grown on naturally labeled vegetable oil and 99 % [$^{13}C_2$] glycine, [$^{13}C_3$] serine, or [^{13}C] formate. Cultivation of *A. gossypii* on naturally labeled medium served as control. Data denote corrected labeling patterns. Data were corrected for fraction of unlabeled biomass in the inoculum as well as occurrence of natural isotopes. The mass isotopomer M+0 represents the relative amount of non-labeled, M+1 the amount of singly-labeled mass isotopomer fraction and so on. Data were obtained from three individual replicates.

Analyte		Control	[$^{13}C_2$] Gly	[^{13}C] For	[$^{13}C_3$] Ser
Ala_260	M+0	0.96 ± 0.00	0.91 ± 0.00	0.91 ± 0.00	0.92 ± 0.00
	M+1	0.03 ± 0.00	0.03 ± 0.00	0.09 ± 0.00	0.04 ± 0.00
	M+2	0.00 ± 0.00	0.06 ± 0.00	0.00 ± 0.00	0.00 ± 0.00
	M+3	0.00 ± 0.00	0.00 ± 0.00	0.00 ± 0.00	0.03 ± 0.00
	SFL [%]	1.15 ± 0.08	5.17 ± 0.11	3.01 ± 0.11	4.89 ± 0.11
	SFL_corr [%]	**0.08 ± 0.01**	**4.22 ± 0.08**	**2.00 ± 0.07**	**4.05 ± 0.09**
Gly_246	M+0	0.98 ± 0.00	0.15 ± 0.06	0.98 ± 0.00	0.89 ± 0.12
	M+1	0.02 ± 0.00	0.02 ± 0.00	0.02 ± 0.00	0.02 ± 0.00
	M+2	0.00 ± 0.00	0.83 ± 0.11	0.00 ± 0.00	0.08 ± 0.01
	SFL [%]	1.15 ± 0.08	84.07 ± 10.64	1.09 ± 0.16	9.44 ± 1.13
	SFL_corr [%]	**0.08 ± 0.01**	**85.41 ± 10.81**	**0.16 ± 0.02**	**8.92 ± 1.07**
Val_288	M+0	0.95 ± 0.00	0.94 ± 0.00	0.94 ± 0.00	0.94 ± 0.00
	M+1	0.05 ± 0.00	0.05 ± 0.00	0.05 ± 0.00	0.05 ± 0.00
	M+2	0.00 ± 0.00	0.00 ± 0.00	0.00 ± 0.00	0.00 ± 0.00
	M+3	0.00 ± 0.00	0.00 ± 0.00	0.00 ± 0.00	0.00 ± 0.00
	M+4	0.00 ± 0.00	0.00 ± 0.00	0.00 ± 0.00	0.00 ± 0.00
	M+5	0.00 ± 0.00	0.00 ± 0.00	0.00 ± 0.00	0.00 ± 0.00
	SFL [%]	1.12 ± 0.03	1.3 ± 0.04	1.24 ± 0.13	1.24 ± 0.11
	SFL_corr [%]	**0.05 ± 0.00**	**0.21 ± 0.01**	**0.15 ± 0.02**	**0.15 ± 0.01**
Leu_302	M+0	0.93 ± 0.00	0.93 ± 0.00	0.93 ± 0.00	0.93 ± 0.00
	M+1	0.07 ± 0.00	0.07 ± 0.00	0.06 ± 0.00	0.07 ± 0.00
	M+2	0.00 ± 0.00	0.00 ± 0.00	0.00 ± 0.00	0.00 ± 0.00
	M+3	0.00 ± 0.00	0.00 ± 0.00	0.00 ± 0.00	0.00 ± 0.00
	M+4	0.00 ± 0.00	0.00 ± 0.00	0.00 ± 0.00	0.00 ± 0.00
	M+5	0.00 ± 0.00	0.00 ± 0.00	0.00 ± 0.00	0.00 ± 0.00
	M+6	0.00 ± 0.00	0.00 ± 0.00	0.00 ± 0.00	0.00 ± 0.00
	SFL [%]	1.16 ± 0.04	1.17 ± 0.01	1.21 ± 0.08	1.21 ± 0.10
	SFL_corr [%]	**0.09 ± 0.00**	**0.07 ± 0.00**	**0.12 ± 0.01**	**0.12 ± 0.01**
Ile_302	M+0	0.93 ± 0.00	0.93 ± 0.00	0.93 ± 0.00	0.93 ± 0.00
	M+1	0.06 ± 0.00	0.07 ± 0.00	0.07 ± 0.00	0.06 ± 0.00
	M+2	0.00 ± 0.00	0.00 ± 0.00	0.00 ± 0.00	0.00 ± 0.00
	M+3	0.00 ± 0.00	0.00 ± 0.00	0.00 ± 0.00	0.00 ± 0.00
	M+4	0.00 ± 0.00	0.00 ± 0.00	0.00 ± 0.00	0.00 ± 0.00
	M+5	0.00 ± 0.00	0.00 ± 0.00	0.00 ± 0.00	0.00 ± 0.00
	M+6	0.00 ± 0.00	0.00 ± 0.00	0.00 ± 0.00	0.00 ± 0.00
	SFL [%]	1.15 ± 0.08	1.19 ± 0.01	1.20 ± 0.03	1.20 ± 0.10
	SFL_corr [%]	**0.05 ± 0.00**	**0.09 ± 0.00**	**0.10 ± 0.00**	**0.11 ± 0.01**

Table 17: Relative mass isotopomer fractions of amino acids from hydrolyzed cell protein of *A. gossypii* B2 grown on naturally labeled vegetable oil and 99 % [$^{13}C_2$] glycine, [$^{13}C_3$] serine, or [^{13}C] formate **(continued)**.

Analyte		Control	[$^{13}C_2$] Gly	[^{13}C] For	[$^{13}C_3$] Ser
Pro_286	M+0	0.94 ± 0.00	0.95 ± 0.00	0.94 ± 0.00	0.95 ± 0.00
	M+1	0.05 ± 0.00	0.05 ± 0.00	0.06 ± 0.00	0.05 ± 0.00
	M+2	0.00 ± 0.00	0.00 ± 0.00	0.00 ± 0.00	0.00 ± 0.00
	M+3	0.00 ± 0.00	0.00 ± 0.00	0.00 ± 0.00	0.00 ± 0.00
	M+4	0.00 ± 0.00	0.00 ± 0.00	0.00 ± 0.00	0.00 ± 0.00
	M+5	0.00 ± 0.00	0.00 ± 0.00	0.00 ± 0.00	0.00 ± 0.00
	SFL [%]	1.09 ± 0.12	1.05 ± 0.05	1.13 ± 0.02	1.13 ± 0.01
	SFL_corr [%]	**0.02 ± 0.00**	**-0.05 ± 0.00**	**0.03 ± 0.00**	**0.03 ± 0.00**
Ser_390	M+0	0.97 ± 0.00	0.37 ± 0.03	0.57 ± 0.02	0.29 ± 0.04
	M+1	0.03 ± 0.00	0.02 ± 0.00	0.43 ± 0.03	0.14 ± 0.02
	M+2	0.00 ± 0.00	0.6 ± 0.03	0.01 ± 0.00	0.04 ± 0.01
	M+3	0.00 ± 0.00	0.01 ± 0.00	0.00 ± 0.00	0.53 ± 0.07
	SFL [%]	1.05 ± 0.11	41.59 ± 2.29	14.68 ± 0.85	60.13 ± 3.61
	SFL_corr [%]	**-0.02 ± 0.00**	**41.67 ± 2.29**	**14.25 ± 0.82**	**63.13 ± 3.79**
Thr_404	M+0	0.96 ± 0.00	0.96 ± 0.00	0.96 ± 0.00	0.95 ± 0.00
	M+1	0.04 ± 0.00	0.04 ± 0.00	0.04 ± 0.00	0.04 ± 0.00
	M+2	0.00 ± 0.00	0.00 ± 0.00	0.00 ± 0.00	0.00 ± 0.00
	M+3	0.00 ± 0.00	0.00 ± 0.00	0.00 ± 0.00	0.00 ± 0.00
	M+4	0.00 ± 0.00	0.00 ± 0.00	0.00 ± 0.00	0.01 ± 0.00
	SFL [%]	1.07 ± 0.03	1.21 ± 0.03	1.04 ± 0.03	1.04 ± 0.01
	SFL_corr [%]	**0.00 ± 0.00**	**0.11 ± 0.00**	**-0.06 ± 0.00**	**-0.06 ± 0.00**
Phe_336	M+0	0.91 ± 0.00	0.90 ± 0.00	0.91 ± 0.00	0.90 ± 0.00
	M+1	0.09 ± 0.00	0.08 ± 0.00	0.09 ± 0.00	0.09 ± 0.00
	M+2	0.00 ± 0.00	0.00 ± 0.00	0.00 ± 0.00	0.00 ± 0.00
	M+3	0.00 ± 0.00	0.00 ± 0.00	0.00 ± 0.00	0.00 ± 0.00
	M+4	0.00 ± 0.00	0.00 ± 0.00	0.00 ± 0.00	0.00 ± 0.00
	M+5	0.00 ± 0.00	0.01 ± 0.00	0.00 ± 0.00	0.00 ± 0.00
	M+6	0.00 ± 0.00	0.00 ± 0.00	0.00 ± 0.00	0.00 ± 0.00
	M+7	0.00 ± 0.00	0.00 ± 0.00	0.00 ± 0.00	0.00 ± 0.00
	M+8	0.00 ± 0.00	0.00 ± 0.00	0.00 ± 0.00	0.00 ± 0.00
	M+9	0.00 ± 0.00	0.00 ± 0.00	0.00 ± 0.00	0.00 ± 0.00
	SFL [%]	1.06 ± 0.03	1.39 ± 0.33	1.06 ± 0.01	1.06 ± 0.03
	SFL_corr [%]	**-0.01 ± 0.00**	**0.30 ± 0.07**	**-0.04 ± 0.00**	**-0.04 ± 0.00**
Asp_418	M+0	0.96 ± 0.00	0.96 ± 0.00	0.96 ± 0.00	0.95 ± 0.00
	M+1	0.04 ± 0.00	0.04 ± 0.00	0.04 ± 0.00	0.04 ± 0.00
	M+2	0.00 ± 0.00	0.00 ± 0.00	0.00 ± 0.00	0.00 ± 0.00
	M+3	0.00 ± 0.00	0.00 ± 0.00	0.00 ± 0.00	0.00 ± 0.00
	M+4	0.00 ± 0.00	0.00 ± 0.00	0.00 ± 0.00	0.00 ± 0.00
	SFL [%]	0.99 ± 0.08	1.17 ± 0.05	1.13 ± 0.10	1.13 ± 0.05
	SFL_corr [%]	**-0.08 ± 0.01**	**0.07 ± 0.00**	**0.03 ± 0.00**	**0.03 ± 0.00**

Table 17: Relative mass isotopomer fractions of amino acids from hydrolyzed cell protein of *A. gossypii* B2 grown on naturally labeled vegetable oil and 99 % [$^{13}C_2$] glycine, [$^{13}C_3$] serine, or [^{13}C] formate **(continued)**.

Analyte		Control	[$^{13}C_2$] Gly	[^{13}C] For	[$^{13}C_3$] Ser
Glu_432	M+0	0.94 ± 0.00	0.94 ± 0.00	0.94 ± 0.00	0.94 ± 0.00
	M+1	0.05 ± 0.00	0.06 ± 0.00	0.05 ± 0.00	0.06 ± 0.00
	M+2	0.00 ± 0.00	0.00 ± 0.00	0.00 ± 0.00	0.00 ± 0.00
	M+3	0.00 ± 0.00	0.00 ± 0.00	0.00 ± 0.00	0.00 ± 0.00
	M+4	0.00 ± 0.00	0.00 ± 0.00	0.00 ± 0.00	0.00 ± 0.00
	M+5	0.00 ± 0.00	0.00 ± 0.00	0.00 ± 0.00	0.00 ± 0.00
	SFL [%]	1.10 ± 0.10	1.30 ± 0.11	1.29 ± 0.10	1.31 ± 0.09
	SFL_corr [%]	**0.03 ± 0.03**	**0.21 ± 0.02**	**0.20 ± 0.02**	**0.22 ± 0.02**
Lys_431	M+0	0.90 ± 0.02	0.94 ± 0.00	0.94 ± 0.00	0.94 ± 0.00
	M+1	0.06 ± 0.00	0.06 ± 0.00	0.05 ± 0.00	0.06 ± 0.00
	M+2	0.04 ± 0.02	0.00 ± 0.00	0.01 ± 0.00	0.00 ± 0.00
	M+3	0.00 ± 0.00	0.00 ± 0.00	0.00 ± 0.00	0.00 ± 0.00
	M+4	0.00 ± 0.00	0.00 ± 0.00	0.00 ± 0.00	0.00 ± 0.00
	M+5	0.00 ± 0.00	0.00 ± 0.00	0.00 ± 0.00	0.00 ± 0.00
	M+6	0.00 ± 0.00	0.00 ± 0.00	0.00 ± 0.00	0.00 ± 0.00
	SFL [%]	1.77 ± 0.68	1.06 ± 0.05	1.06 ± 0.09	1.06 ± 0.01
	SFL_corr [%]	**0.70 ± 0.27**	**-0.04 ± 0.00**	**-0.04 ± 0.00**	**-0.04 ± 0.00**
Arg_442	M+0	0.94 ± 0.00	0.94 ± 0.00	0.93 ± 0.00	0.93 ± 0.00
	M+1	0.06 ± 0.00	0.06 ± 0.00	0.07 ± 0.00	0.06 ± 0.00
	M+2	0.00 ± 0.00	0.00 ± 0.00	0.00 ± 0.00	0.00 ± 0.00
	M+3	0.00 ± 0.00	0.00 ± 0.00	0.00 ± 0.00	0.00 ± 0.00
	M+4	0.00 ± 0.00	0.00 ± 0.00	0.00 ± 0.00	0.00 ± 0.00
	M+5	0.00 ± 0.00	0.00 ± 0.00	0.00 ± 0.00	0.00 ± 0.00
	M+6	0.00 ± 0.00	0.00 ± 0.00	0.00 ± 0.00	0.00 ± 0.00
	SFL [%]	1.12 ± 0.06	1.08 ± 0.06	1.11 ± 0.09	1.11 ± 0.09
	SFL_corr [%]	**0.05 ± 0.00**	**-0.02 ± 0.00**	**0.01 ± 0.00**	**0.01 ± 0.00**
Tyr_466	M+0	0.92 ± 0.00	0.90 ± 0.00	0.91 ± 0.00	0.90 ± 0.00
	M+1	0.07 ± 0.00	0.09 ± 0.00	0.09 ± 0.00	0.09 ± 0.00
	M+2	0.00 ± 0.00	0.01 ± 0.00	0.00 ± 0.00	0.01 ± 0.00
	M+3	0.00 ± 0.00	0.00 ± 0.00	0.00 ± 0.00	0.00 ± 0.00
	M+4	0.00 ± 0.00	0.00 ± 0.00	0.00 ± 0.00	0.00 ± 0.00
	M+5	0.00 ± 0.00	0.00 ± 0.00	0.00 ± 0.00	0.00 ± 0.00
	M+6	0.00 ± 0.00	0.00 ± 0.00	0.00 ± 0.00	0.00 ± 0.00
	M+7	0.00 ± 0.00	0.00 ± 0.00	0.00 ± 0.00	0.00 ± 0.00
	M+8	0.00 ± 0.00	0.00 ± 0.00	0.00 ± 0.00	0.00 ± 0.00
	M+9	0.00 ± 0.00	0.00 ± 0.00	0.00 ± 0.00	0.00 ± 0.00
	SFL [%]	1.00 ± 0.03	1.27 ± 0.16	1.08 ± 0.03	1.08 ± 0.07
	SFL_corr [%]	**-0.07 ± 0.00**	**0.17 ± 0.02**	**-0.02 ± 0.00**	**-0.02 ± 0.00**

Table 18: Relative mass isotopomer fractions of amino acids from hydrolyzed cell protein of *A. gossypii* B2 grown on naturally labeled vegetable oil and 99 % [$U^{13}C$] yeast extract, [$^{13}C_5$] glutamate, or the combined addition of [$^{13}C_2$] glycine, [^{13}C] formate, [$^{13}C_5$] glutamate, and [$U^{13}C$] yeast extract. Cultivation of *A. gossypii* on naturally labeled medium served as control. Data denote corrected labeling patterns. Data were corrected for fraction of unlabeled biomass in the inoculum as well as occurrence of natural isotopes. The mass isotopomer M+0 represents the relative amount of non-labeled, M+1 the amount of singly-labeled mass isotopomer fraction and so on. Data were obtained from three individual replicates.

Analyte		Control	[$U^{13}C$] YE	[$^{13}C_5$] Glu	[$U^{13}C$] Gly, For, Glu, YE
Ala_260	M+0	0.97 ± 0.00	0.31 ± 0.02	0.94 ± 0.05	0.22 ± 0.03
	M+1	0.03 ± 0.00	0.05 ± 0.00	0.05 ± 0.00	0.04 ± 0.00
	M+2	0.00 ± 0.00	0.04 ± 0.00	0.01 ± 0.00	0.04 ± 0.00
	M+3	0.00 ± 0.00	0.61 ± 0.06	0.01 ± 0.00	0.7 ± 0.08
	SFL [%]	1.13 ± 0.05	65.06 ± 4.01	2.71 ± 0.17	73.96 ± 3.04
	SFL_corr [%]	**0.03 ± 0.00**	**71.89 ± 4.43**	**1.69 ± 0.04**	**81.89 ± 3.37**
Gly_246	M+0	0.98 ± 0.00	0.86 ± 0.09	0.98 ± 0.00	0.01 ± 0.00
	M+1	0.02 ± 0.00	0.02 ± 0.00	0.02 ± 0.00	0.02 ± 0.00
	M+2	0.00 ± 0.00	0.12 ± 0.01	0.00 ± 0.00	0.97 ± 0.00
	SFL [%]	1.11 ± 0.04	12.95 ± 1.25	1.23 ± 0.05	97.89 ± 2.64
	SFL_corr [%]	**0.01 ± 0.00**	**12.20 ± 1.18**	**0.13 ± 0.01**	**99.61 ± 2.69**
Val_288	M+0	0.95 ± 0.00	0.07 ± 0.00	0.95 ± 0.00	0.07 ± 0.00
	M+1	0.05 ± 0.00	0.01 ± 0.00	0.05 ± 0.00	0.01 ± 0.00
	M+2	0.00 ± 0.00	0.01 ± 0.00	0.00 ± 0.00	0.01 ± 0.00
	M+3	0.00 ± 0.00	0.01 ± 0.00	0.00 ± 0.00	0.01 ± 0.00
	M+4	0.00 ± 0.00	0.05 ± 0.01	0.00 ± 0.00	0.05 ± 0.00
	M+5	0.00 ± 0.00	0.86 ± 0.8	0.00 ± 0.00	0.86 ± 0.9
	SFL [%]	1.09 ± 0.03	91.05 ± 7.50	1.15 ± 0.02	90.89 ± 9.01
	SFL_corr [%]	**-0.01 ± 0.00**	**103.92 ± 8.56**	**0.05 ± 0.01**	**103.74 ± 10.28**
Leu_274	M+0	0.95 ± 0.00	0.06 ± 0.00	0.95 ± 0.00	0.07 ± 0.00
	M+1	0.05 ± 0.00	0.00 ± 0.00	0.05 ± 0.00	0.00 ± 0.00
	M+2	0.00 ± 0.00	0.00 ± 0.00	0.00 ± 0.00	0.00 ± 0.00
	M+3	0.00 ± 0.00	0.00 ± 0.00	0.00 ± 0.00	0.00 ± 0.00
	M+4	0.00 ± 0.00	0.05 ± 0.00	0.00 ± 0.00	0.05 ± 0.00
	M+5	0.00 ± 0.00	0.88 ± 0.10	0.00 ± 0.00	0.87 ± 0.08
	SFL [%]	1.09 ± 0.01	92.16 ± 4.36	1.11 ± 0.02	91.42 ± 5.47
	SFL_corr [%]	**-0.01 ± 0.00**	**104.33 ± 4.94**	**0.05 ± 0.01**	**103.48 ± 6.19**
Ile_274	M+0	0.94 ± 0.00	0.09 ± 0.01	0.93 ± 0.07	0.09 ± 0.02
	M+1	0.06 ± 0.00	0.01 ± 0.00	0.06 ± 0.00	0.01 ± 0.00
	M+2	0.00 ± 0.00	0.04 ± 0.00	0.00 ± 0.00	0.04 ± 0.00
	M+3	0.00 ± 0.00	0.01 ± 0.00	0.00 ± 0.00	0.01 ± 0.00
	M+4	0.00 ± 0.00	0.05 ± 0.00	0.00 ± 0.00	0.05 ± 0.00
	M+5	0.00 ± 0.00	0.8 ± 0.07	0.00 ± 0.00	0.79 ± 0.08
	SFL [%]	1.24 ± 0.12	86.35 ± 6.53	1.39 ± 0.21	86.08 ± 10.11
	SFL_corr [%]	**0.14 ± 0.01**	**98.87 ± 7.48**	**0.10 ± 0.02**	**98.56 ± 11.58**

Table 18: Relative mass isotopomer fractions of amino acids from hydrolyzed cell protein of *A. gossypii* B2 grown on naturally labeled vegetable oil and 99 % [U^{13}C] yeast extract, [^{13}C$_5$] glutamate, or the combined addition of [^{13}C$_2$] glycine, [^{13}C] formate, [^{13}C$_5$] glutamate, and [U^{13}C] yeast extract **(continued)**.

Analyte		Control	[U^{13}C] YE	[^{13}C$_5$] Glu	[U^{13}C] Gly, For, Glu, YE
Pro_258	M+0	0.95 ± 0.00	n.d.	0.85 ± 0.09	0.26 ± 0.03
	M+1	0.05 ± 0.00	n.d.	0.08 ± 0.02	0.06 ± 0.00
	M+2	0.00 ± 0.00	n.d.	0.02 ± 0.00	0.03 ± 0.00
	M+3	0.00 ± 0.00	n.d.	0.00 ± 0.00	0.03 ± 0.00
	M+4	0.00 ± 0.00	n.d.	0.04 ± 0.00	0.62 ± 0.05
	SFL [%]	1.17 ± 0.10	n.d.	7.59 ± 0.60	67.25 ± 4.02
	SFL$_{corr}$ [%]	**0.07 ± 0.01**	**n.d.**	**6.81 ± 0.54**	**76.00 ± 4.54**
Ser_390	M+0	0.97 ± 0.00	0.58 ± 0.04	0.96 ± 0.02	0.01 ± 0.00
	M+1	0.03 ± 0.00	0.17 ± 0.02	0.04 ± 0.00	0.02 ± 0.00
	M+2	0.00 ± 0.00	0.08 ± 0.00	0.00 ± 0.00	0.11 ± 0.01
	M+3	0.00 ± 0.00	0.17 ± 0.01	0.00 ± 0.00	0.87 ± 0.09
	SFL [%]	1.08 ± 0.05	28.01 ± 1.32	1.33 ± 0.11	94.47 ± 6.78
	SFL$_{corr}$ [%]	**-0.02 ± 0.01**	**31.52 ± 1.49**	**0.24 ± 0.02**	**109.38 ± 7.85**
Thr_404	M+0	0.96 ± 0.00	0.16 ± 0.02	0.94 ± 0.03	0.13 ± 0.02
	M+1	0.04 ± 0.00	0.02 ± 0.00	0.05 ± 0.00	0.02 ± 0.00
	M+2	0.00 ± 0.00	0.01 ± 0.00	0.01 ± 0.00	0.01 ± 0.00
	M+3	0.00 ± 0.00	0.04 ± 0.00	0.00 ± 0.00	0.04 ± 0.00
	M+4	0.00 ± 0.00	0.77 ± 0.06	0.01 ± 0.00	0.79 ± 0.08
	SFL [%]	1.09 ± 0.03	81.01 ± 7.98	2.26 ± 0.35	83.14 ± 4.41
	SFL$_{corr}$ [%]	**-0.01 ± 0.00**	**87.60 ± 8.63**	**1.22 ± 0.19**	**89.93 ± 4.77**
Phe_336	M+0	0.91 ± 0.00	0.06 ± 0.00	0.91 ± 0.00	0.06 ± 0.00
	M+1	0.08 ± 0.00	0.01 ± 0.00	0.09 ± 0.00	0.01 ± 0.00
	M+2	0.00 ± 0.00	0.00 ± 0.00	0.00 ± 0.00	0.00 ± 0.00
	M+3	0.00 ± 0.00	0.00 ± 0.00	0.00 ± 0.00	0.00 ± 0.00
	M+4	0.00 ± 0.00	0.00 ± 0.00	0.00 ± 0.00	0.00 ± 0.00
	M+5	0.00 ± 0.00	0.00 ± 0.00	0.00 ± 0.00	0.00 ± 0.00
	M+6	0.00 ± 0.00	0.00 ± 0.00	0.00 ± 0.00	0.00 ± 0.00
	M+7	0.00 ± 0.00	0.01 ± 0.00	0.00 ± 0.00	0.01 ± 0.00
	M+8	0.00 ± 0.00	0.08 ± 0.00	0.00 ± 0.00	0.08 ± 0.00
	M+9	0.00 ± 0.00	0.85 ± 0.09	0.00 ± 0.00	0.84 ± 0.08
	SFL [%]	1.14 ± 0.07	92.29 ± 6.52	1.05 ± 0.02	91.9 ± 8.71
	SFL$_{corr}$ [%]	**0.04 ± 0.00**	**104.11 ± 6.99**	**0.00 ± 0.00**	**103.67 ± 9.83**
Asp_418	M+0	0.96 ± 0.00	0.56 ± 0.05	0.87 ± 0.08	0.49 ± 0.06
	M+1	0.04 ± 0.00	0.07 ± 0.00	0.07 ± 0.00	0.10 ± 0.01
	M+2	0.00 ± 0.00	0.04 ± 0.00	0.03 ± 0.00	0.06 ± 0.00
	M+3	0.00 ± 0.00	0.02 ± 0.00	0.00 ± 0.00	0.03 ± 0.00
	M+4	0.00 ± 0.00	0.3 ± 0.02	0.02 ± 0.00	0.32 ± 0.04
	SFL [%]	1.00 ± 0.08	35.62 ± 2.44	5.6 ± 0.52	39.72 ± 2.02
	SFL$_{corr}$ [%]	**-0.10 ± 0.01**	**43.11 ± 2.95**	**4.71 ± 0.44**	**48.23 ± 2.45**

Table 18: Relative mass isotopomer fractions of amino acids from hydrolyzed cell protein of *A. gossypii* B2 grown on naturally labeled vegetable oil and 99 % $[U^{13}C]$ yeast extract, $[^{13}C_5]$ glutamate, or the combined addition of $[^{13}C_2]$ glycine, $[^{13}C]$ formate, $[^{13}C_5]$ glutamate, and $[U^{13}C]$ yeast extract **(continued)**.

Analyte		Control	$[U^{13}C]$ YE	$[^{13}C_5]$ Glu	$[U^{13}C]$ Gly, For, Glu, YE
Glu_432	M+0	0.95 ± 0.00	0.73 ± 0.07	0.78 ± 0.09	0.57 ± 0.07
	M+1	0.05 ± 0.00	0.10 ± 0.01	0.09 ± 0.02	0.13 ± 0.02
	M+2	0.00 ± 0.00	0.05 ± 0.00	0.02 ± 0.00	0.06 ± 0.00
	M+3	0.00 ± 0.00	0.03 ± 0.00	0.03 ± 0.00	0.05 ± 0.00
	M+4	0.00 ± 0.00	0.01 ± 0.00	0.00 ± 0.00	0.01 ± 0.00
	M+5	0.00 ± 0.00	0.09 ± 0.01	0.08 ± 0.00	0.17 ± 0.01
	SFL [%]	1.03 ± 0.04	14.78 ± 1.12	12.32 ± 1.31	26.11 ± 1.03
	SFL_corr [%]	**-0.07 ± 0.00**	**14.33 ± 1.09**	**11.76 ± 1.25**	**26.21 ± 1.03**
Lys_431	M+0	0.94 ± 0.00	0.07 ± 0.00	0.94 ± 0.00	0.07 ± 0.00
	M+1	0.05 ± 0.00	0.00 ± 0.00	0.06 ± 0.01	0.00 ± 0.00
	M+2	0.00 ± 0.00	0.00 ± 0.00	0.00 ± 0.00	0.00 ± 0.00
	M+3	0.00 ± 0.00	0.00 ± 0.00	0.00 ± 0.00	0.00 ± 0.00
	M+4	0.00 ± 0.00	0.04 ± 0.00	0.00 ± 0.00	0.02 ± 0.00
	M+5	0.00 ± 0.00	0.07 ± 0.00	0.00 ± 0.00	0.06 ± 0.00
	M+6	0.00 ± 0.00	0.81 ± 0.09	0.00 ± 0.00	0.84 ± 0.09
	SFL [%]	0.98 ± 0.12	89.61 ± 9.01	0.94 ± 0.20	90.68 ± 8.69
	SFL_corr [%]	**-0.12 ± 0.01**	**110.19 ± 11.08**	**0.00 ± 0.00**	**111.51 ± 10.69**
Arg_442	M+0	0.94 ± 0.00	0.14 ± 0.01	0.85 ± 0.08	0.08 ± 0.00
	M+1	0.05 ± 0.00	0.03 ± 0.00	0.09 ± 0.01	0.03 ± 0.00
	M+2	0.00 ± 0.00	0.01 ± 0.00	0.01 ± 0.00	0.01 ± 0.00
	M+3	0.00 ± 0.00	0.00 ± 0.00	0.01 ± 0.00	0.01 ± 0.00
	M+4	0.00 ± 0.00	0.01 ± 0.00	0.00 ± 0.00	0.01 ± 0.00
	M+5	0.00 ± 0.00	0.08 ± 0.00	0.02 ± 0.00	0.08 ± 0.01
	M+6	0.00 ± 0.00	0.73 ± 0.08	0.00 ± 0.00	0.79 ± 0.08
	SFL [%]	1.10 ± 0.04	80.83 ± 4.74	5.04 ± 0.67	87.24 ± 10.41
	SFL_corr [%]	**0.00 ± 0.00**	**83.87 ± 4.92**	**4.13 ± 0.55**	**90.61 ± 10.81**
Tyr_466	M+0	0.92 ± 0.00	0.07 ± 0.00	0.8 ± 0.09	0.06 ± 0.00
	M+1	0.08 ± 0.00	0.02 ± 0.00	0.13 ± 0.03	0.02 ± 0.00
	M+2	0.00 ± 0.00	0.01 ± 0.00	0.04 ± 0.00	0.01 ± 0.00
	M+3	0.00 ± 0.00	0.01 ± 0.00	0.02 ± 0.00	0.01 ± 0.00
	M+4	0.00 ± 0.00	0.00 ± 0.00	0.01 ± 0.00	0.00 ± 0.00
	M+5	0.00 ± 0.00	0.00 ± 0.00	0.00 ± 0.00	0.00 ± 0.00
	M+6	0.00 ± 0.00	0.00 ± 0.00	0.00 ± 0.00	0.00 ± 0.00
	M+7	0.00 ± 0.00	0.01 ± 0.00	0.00 ± 0.00	0.01 ± 0.00
	M+8	0.00 ± 0.00	0.08 ± 0.01	0.00 ± 0.00	0.07 ± 0.00
	M+9	0.00 ± 0.00	0.80 ± 0.07	0.00 ± 0.00	0.81 ± 0.09
	SFL [%]	1.14 ± 0.05	88.17 ± 6.35	3.45 ± 0.31	89.8 ± 11.04
	SFL_corr [%]	**0.04 ± 0.00**	**92.08 ± 6.63**	**2.47 ± 0.22**	**93.80 ± 11.53**

Table 18: Relative mass isotopomer fractions of amino acids from hydrolyzed cell protein of *A. gossypii* B2 grown on naturally labeled vegetable oil and 99 % [U^{13}C] yeast extract, [^{13}C$_5$] glutamate, or the combined addition of [^{13}C$_2$] glycine, [^{13}C] formate, [^{13}C$_5$] glutamate, and [U^{13}C] yeast extract **(continued)**.

Analyte		Control	[U^{13}C] YE	[^{13}C$_5$] Glu	[U^{13}C] Gly, For, Glu, YE
His_440	M+0	0.93 ± 0.00	0.07 ± 0.00	0.93 ± 0.00	n.d.
	M+1	0.06 ± 0.00	0.05 ± 0.00	0.06 ± 0.00	n.d.
	M+2	0.00 ± 0.00	0.01 ± 0.00	0.00 ± 0.00	n.d.
	M+3	0.00 ± 0.00	0.01 ± 0.00	0.01 ± 0.00	n.d.
	M+4	0.00 ± 0.00	0.01 ± 0.00	0.00 ± 0.00	n.d.
	M+5	0.00 ± 0.00	0.06 ± 0.02	0.00 ± 0.00	n.d.
	M+6	0.00 ± 0.00	0.80 ± 0.10	0.00 ± 0.00	n.d.
	SFL [%]	1.21 ± 0.13	86.75 ± 4.01	1.43 ± 0.35	n.d.
	SFL$_{corr}$ [%]	**0.11 ± 0.01**	**103.64 ± 4.79**	**0.34 ± 0.08**	**n.d.**
Glc_554	M+0	0.94 ± 0.00	0.73 ± 0.07	0.74 ± 0.08	0.55 ± 0.06
	M+1	0.05 ± 0.00	0.17 ± 0.02	0.16 ± 0.02	0.23 ± 0.02
	M+2	0.00 ± 0.00	0.05 ± 0.00	0.05 ± 0.00	0.11 ± 0.01
	M+3	0.00 ± 0.00	0.04 ± 0.00	0.04 ± 0.00	0.09 ± 0.01
	M+4	0.00 ± 0.00	0.01 ± 0.00	0.00 ± 0.00	0.02 ± 0.00
	M+5	0.00 ± 0.00	0.00 ± 0.00	0.00 ± 0.00	0.01 ± 0.00
	M+6	0.00 ± 0.00	0.00 ± 0.00	0.00 ± 0.00	0.00 ± 0.00
	SFL [%]	1.14 ± 0.10	7.24 ± 0.50	7.12 ± 0.39	13.78 ± 1.52
	SFL$_{corr}$ [%]	**0.04 ± 0.00**	**6.20 ± 0.43**	**6.30 ± 0.35**	**12.81 ± 1.41**

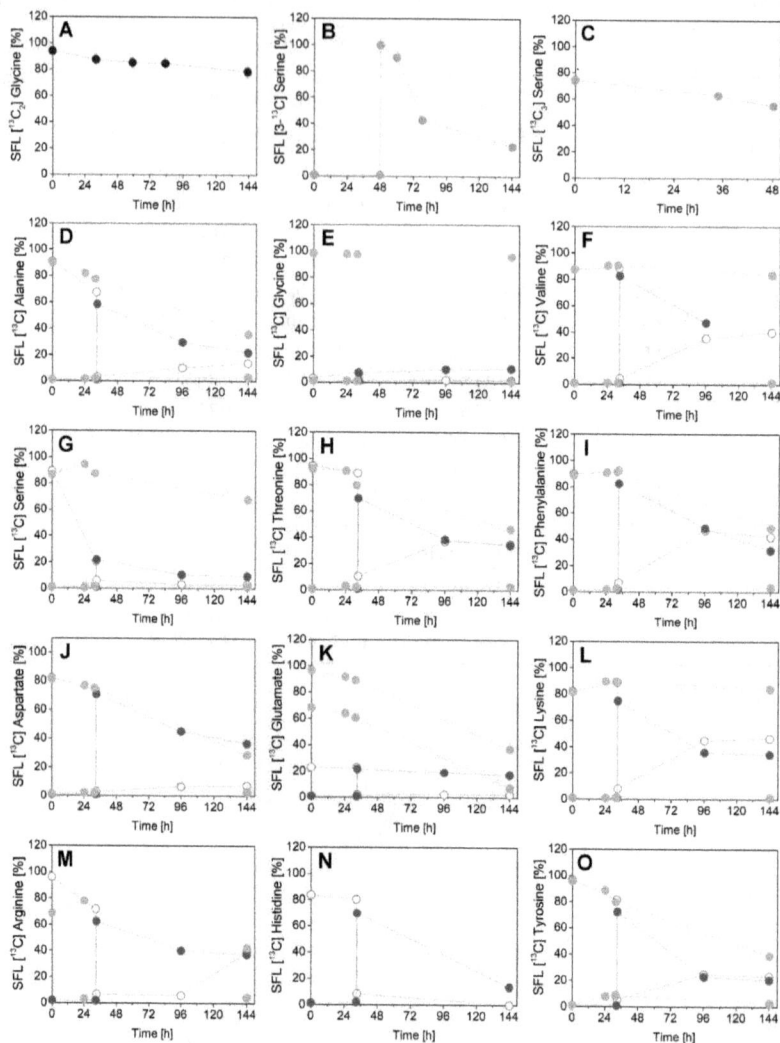

Figure 40: Change in summed fractional labeling (SFL) in the culture supernatant for various experiments throughout the course of the cultivation: cultivation of *A. gossypii* B2 on complex medium with naturally labeled rapeseed oil and [$^{13}C_2$] glycine (A) (dark blue), [3-^{13}C] serine after 48 h (B) (green), [$^{13}C_3$] serine (C) (green), [U^{13}C] yeast extract added from 0-32 h (D-O) (open circles), [U^{13}C] yeast extract added from 32-144 h (D-O) (purple), [$^{13}C_5$] glutamate (D-O) (light blue), and a combination of [U^{13}C] yeast extract, [$^{13}C_2$] glycine, [$^{13}C_5$] glutamate, and [^{13}C] formate (D-O) (grey). Unless otherwise indicated, tracers were added at the beginning of the cultivation. For experiments with labeled serine and glycine (without other labeled compounds), only the ^{13}C labeling of the respective tracer was measured in the supernatant. For experiments with glutamate and yeast extract several amino acids were measured regarding ^{13}C enrichment (D-O). Labeling information of serine (B) refers to the terminal carbon atom only and 33 % of labeling were set to 100 %. Data denote mean values from three individual replicates with a mean standard deviation of 5 %.

6.5 Biomass composition of *A. gossypii* for flux calculations

Table 19: Anabolic precursor demand of *A. gossypii* for biomass synthesis derived from the genome-scale metabolic model (Ledesma-Amaro et al., 2014a), adjusted for the content and composition of lipids for growth on vegetable oil (Stahmann et al., 1994) and underlying pathway stoichiometry for riboflavin production during the growth phase (Bacher et al., 2000; Kanehisa et al., 2017; Kanehisa and Goto, 2000; Kanehisa et al., 2016). 3PG, 3-phosphoglycerate; AcCoA, acetyl-CoA; AKG, α-ketoglutarate; G3P, glyceraldehyde 3-phosphate; G6P, glucose 6-phosphate; E4P, erythrose 4-phosphate; F6P, fructose 6-phosphate; OAA, oxaloacetate; PEP, phosphoenolpyruvate; PYR, pyruvate; R5P, ribose 5-phosphate.

Precursor	Demand [μmol g⁻¹]	G6P	F6P	R5P	E4P	G3P	3PG	PEP	PYR	AcCoA	OAA	AKG
Alanine	357.3								1			
Arginine	135.8											1
Asparagine	171.5										1	
Aspartate	171.5										1	
Cysteine	42.9						1					
Glutamine	268.0											1
Glutamate	268.0											1
Glycine	325.2						1					
Histidine	75.0			1								
Isoleucine	171.5								1		1	
Leucine	250.1								2	1		
Lysine	239.4								1		1	
Methionine	50.0										1	
Phenylalanine	114.3				1			2				
Proline	128.6											1
Serine	253.7						1					
Threonine	196.5										1	
Tryptophan	28.0			1	1			1				
Tyrosine	96.5				1			2				
Valine	257.3								2			
Protein		0	0	103	239	0	622	450	1783	250	1001	800
ATP	51.0			1			1					
GTP	20.0			1			1					
CTP	51.0			1							1	
UTP	67.0			1							1	
RNA		0	0	189	0	0	71	0	0	0	118	0
dATP	3.6			1			1					
dGTP	2.4			1			1					
dCTP	2.4			1							1	
dTTP	3.6			1							1	
DNA		0	0	12	0	0	6	0	0	0	6	0
Lipid	240.4					1				26		
Glycogen	581.5	1										
Mannan	821.0		1									
Trehalose	23.3	1										
Riboflavin	8.6			3			1					
Total		605	821	330	239	240	707	450	1783	6573	1125	800

Table 20: Contribution of nutrient uptake from the medium and *de novo* synthesis of precursors to the supply of cellular building blocks for *A. gossypii*. Values for the precursor demand were taken from Ledesma-Amaro et al. (2014a) and adjusted for growth on vegetable oil based on Stahmann et al. (1994). Correlation of the demand values with experimental summed fractional labeling (SFL) data from combined results of parallel ^{13}C isotope studies (appendix Table 17 and Table 18) yielded the *de novo* precursor demand for growth on complex medium and rapeseed oil. The full length bar indicates the total precursor demand: the purple fraction depicts the measured percentage taken up from the medium, while the grey fraction depicts the resulting *de novo* biosynthetic fraction of the precursor. Since the ^{13}C labeling could not be measured for all metabolites, following assumptions were made. For the amino acids cysteine, methionine, and tryptophan 100 % uptake were defined, since measured amino acids with a complex biosynthesis were taken up from the medium (e.g. phenylalanine, isoleucine). Since the purine and pyrimidine biosyntheses are feedback regulated pathways (Burns, 1964; Lacroute, 1968; Rolfes, 2006), the presence of nucleotides in yeast extract has been reported (Zhang et al., 2003) and could also be measured in riboflavin samples even at a late stage of cultivation (after 144 h), uptake of DNA and RNA building blocks as well as the GTP-precursor for riboflavin (Figure 32) was defined as 100 %. For trehalose the same ratio was assumed as for the glycogen pool, since the two building blocks share the same precursor glucose 6-phosphate. For the flux calculations lipids were considered to be completely *de novo* synthesized, however the more likely scenario is that fatty acids are taken up by the cell and re-esterified with glycerol inside the cell. Stahmann et al. (1994) reported that the lipid composition of the substrate oil resembled the lipid composition of storage triacylglycerides in *A. gossypii* cells. Due to that, the acetyl-CoA *de novo* demand is not specified in this table. ATP, adenosine triphosphate; CTP, cytidine triphosphate; dATP, deoxyadenosine triphosphate; dCTP, deoxycytidine triphosphate; dGTP, deoxyguanosine triphosphate; dTTP, deoxythymidine triphosphate; GTP, guanosine triphosphate; UTP, uridine triphosphate.

Precursor	Total demand		Uptake	De novo biosynthesis	Resulting de novo demand
	[µmol g^{-1}]		[%]	[%]	[µmol g^{-1}]
Alanine	357.3		80 ± 5	20 ± 5	71.5 ± 15.4
Arginine	135.8		90 ± 4	10 ± 4	13.6 ± 5.4
Asparagine	171.5		89 ± 9	11 ± 9	18.9 ± 14.3
Aspartate	171.5		48 ± 4	52 ± 4	89.4 ± 5.2
Cysteine‡	42.9		100	0	0.0
Glutamine	268.0		74 ± 4	26 ± 4	69.7 ± 9.4
Glutamate	268.0		27 ± 2	74 ± 2	197.0 ± 4.0
Glycine	325.2		98 ± 1	2 ± 1	6.5 ± 30.5
Histidine	75.0		100 ± 5	0 ± 5	0.0 ± 0.0
Isoleucine	171.5		100 ± 7	0 ± 7	0.0 ± 0.0
Leucine	250.1		100 ± 4	0 ± 4	0.0 ± 0.0
Lysine	239.4		100 ± 11	0 ± 11	0.0 ± 0.1
Methionine	50.0		100	0	0.0
Phenylalanine	114.3		100 ± 7	0 ± 7	0.0
Proline	128.6		74 ± 4	26 ± 4	33.4 ± 4.8
Serine	253.7		88 ± 3	12 ± 3	30.4 ± 8.0
Threonine	196.5		89 ± 9	11 ± 9	21.6 ± 17.5
Tryptophan‡	28.0		100	0	0.0
Tyrosine	96.5		95 ± 7	5 ± 7	4.8 ± 6.7
Valine	257.3		100 ± 9	0 ± 9	0.0 ± 0.0
ATP‡	51.0		100	0	0.0
GTP‡	20.0		100	0	0.0
CTP‡	51.0		100	0	0.0
UTP‡	67.0		100	0	0.0
dATP‡	3.6		100	0	0.0
dGTP‡	2.4		100	0	0.0
dCTP‡	2.4		100	0	0.0
dTTP‡	3.6		100	0	0.0
Lipid	240.4		n.d.	n.d.	n.d.
Glycogen	581.5		13 ± 0	88 ± 0	508.8 ± 2.6
Mannan‡	821.0		0	100	821.0
Trehalose	23.3		13 ± 0	88 ± 0	20.4 ± 0.0
Riboflavin‡	8.6		53	47	4.0

‡ The demand for the precursor is assumed. Therefore, no standard deviation could be calculated for the according values.

6.6 Determination of measured fluxes and metabolite balances for flux calculations during growth of *A. gossypii*

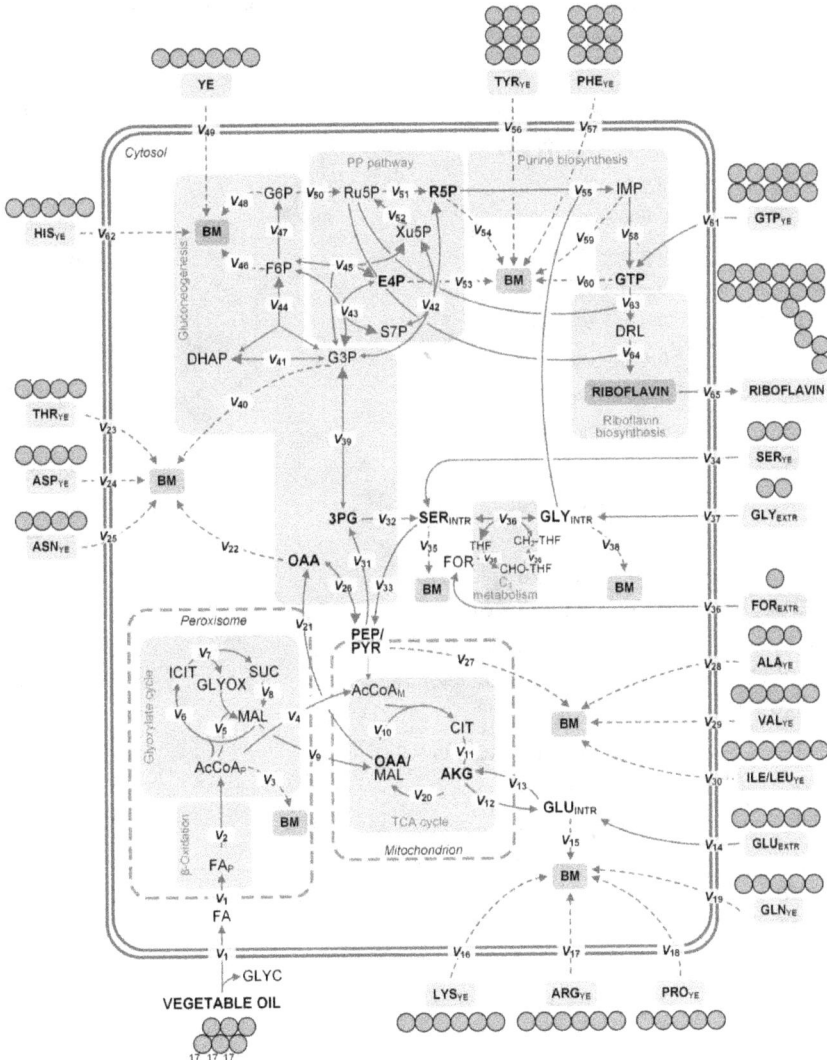

Figure 41: Metabolic network of *A. gossypii*, including extracellular reactions, reactions between intermediary metabolite pools, and anabolic reactions. The reactions numbers refer to the formulation of metabolite balances (see below). The direction of net reactions is indicated by size of arrow head. For abbreviations see Figure 27.

127

In order to calculate carbon fluxes during the growth phase of *A. gossypii*, the following balances were formulated for the network, depicted in Figure 41, and represent growth on vegetable oil and yeast extract. Metabolite balances were expressed, using the numbering of the reactions as presented in Figure 41.

FA_P	$0 = 26\,v_1 - v_2$	(Eq. 15)
$AcCoA_P$	$0 = v_2 - v_3 - v_4 - v_5 - v_6$	(Eq. 16)
ICIT	$0 = v_6 - v_7$	(Eq. 17)
GLYOX	$0 = v_7 - v_5$	(Eq. 18)
SUC	$0 = v_7 - v_8$	(Eq. 19)
MAL	$0 = v_5 + v_8 - v_6 - v_9$	(Eq. 20)
OAA/MAL	$0 = v_9 + v_{20} - v_{10} - v_{21}$	(Eq. 21)
$AcCoA_M$	$0 = v_4 - v_{10}$	(Eq. 22)
CIT	$0 = v_{10} - v_{11}$	(Eq. 23)
AKG	$0 = v_{11} + v_{13} - v_{12} - v_{20}$	(Eq. 24)
GLU_{INTR}	$0 = v_{12} + v_{14} - v_{13} - v_{15}$	(Eq. 25)
OAA	$0 = v_{21} - v_{22} - v_{26}$	(Eq. 26)
PEP/PYR	$0 = v_{26} + v_{33} - v_{27} - v_{31}$	(Eq. 27)
3PG	$0 = v_{31} - v_{32} - v_{39}$	(Eq. 28)
SER_{INTR}	$0 = v_{32} + v_{34} + v_{36} - v_{33} - v_{35}$	(Eq. 29)
GLY_{INTR}	$0 = v_{37} - v_{36} - v_{38} - v_{55}$	(Eq. 30)
G3P	$0 = v_{39} + v_{43} - v_{40} - v_{41} - v_{42} - v_{44} - v_{45}$	(Eq. 31)
DHAP	$0 = v_{41} - v_{44}$	(Eq. 32)
F6P	$0 = v_{44} - v_{43} - v_{45} - v_{46} - v_{47}$	(Eq. 33)
S7P	$0 = v_{43} - v_{42}$	(Eq. 34)
E4P	$0 = v_{45} - v_{43} - v_{53}$	(Eq. 35)
G6P	$0 = v_{47} - v_{48} - v_{50}$	(Eq. 36)
Ru5P	$0 = v_{50} + v_{52} - v_{51} - v_{63} - v_{64}$	(Eq. 37)
Xu5P	$0 = v_{42} + v_{45} - v_{52}$	(Eq. 38)
R5P	$0 = v_{42} + v_{51} - v_{54} - v_{55}$	(Eq. 39)
IMP	$0 = v_{55} - v_{58} - v_{59}$	(Eq. 40)
GTP	$0 = v_{58} + v_{61} - v_{60} - v_{63}$	(Eq. 41)
DRL	$0 = v_{63} - v_{64}$	(Eq. 42)
RF	$0 = v_{64} - v_{65}$	(Eq. 43)

The stoichiometric factor for v_1 in Equation (15) is derived from the assumption that one molecule oil is composed of one molecule glycerol and three fatty acid chains with a chain length of 17.3 carbon atoms each. Thus, all three chains would contribute 52 carbon atoms, which are degraded to 26 molecules acetyl-CoA. The rank of the stoichiometric matrix, formulated for the Equations (15) to (43) equaled 29. This confirmed that the 29 metabolite balances were linearly independent. Since the complete network comprised a total of 65 fluxes, additional 36 pieces of information were required in addition to the stoichiometric balances. Anabolic fluxes based on the biomass composition of *A. gossypii* (Table 19), which were correlated with the amount of precursor taken up by the cell versus the amount of *de novo* biosynthesis (i.e. ^{13}C labeling information, Table 19, Table 20), delivered another 28 measured fluxes (v_3, v_{15}, v_{16}, v_{17}, v_{18}; v_{19}, v_{22}, v_{23}, v_{24}, v_{25}, v_{27}, v_{28}, v_{29}, v_{30}, v_{35}, v_{38}, v_{40}, v_{46}, v_{48}; v_{49}, v_{53}, v_{54}, v_{56}, v_{57}, v_{59}, v_{60}, v_{61}, v_{62}). The remaining 8 pieces of information were obtained through the measurement of extracellular fluxes, i.e. substrate uptake rates (v_1, v_{14}, v_{37}) and riboflavin secretion rate (v_{65}), and the measurement of intracellular fluxes derived from additional ^{13}C labeling information (v_{12}, v_{33}, v_{34}, v_{36}) (Table 17, Table 18). Anabolic fluxes were calculated by multiplying the anabolic demand with the biomass yield, $Y_{X/Oil}$, and the uptake rate of the main carbon source, $q_{S,Oil}$ (Table 6). In total, 65 pieces of information were obtained, which rendered a fully determined metabolic network.

6.7   Data from LC/MS analyses

Table 21: Relative mass isotopomer fractions of riboflavin (*m/z* 377) from culture supernatants of *A. gossypii* WT and B2 grown on naturally labeled complex medium and [$^{13}C_6$] glucose. The control represents the labeling data for *A. gossypii* grown on fully naturally labeled medium. Data denote experimental LC/MS data (exp) and corrected labeling patterns (corr). Data were corrected for occurrence of natural isotopes. The mass isotopomer M+0 represents the relative amount of non-labeled, M+1 the amount of singly-labeled mass isotopomer fraction and so on. Data were obtained from three individual replicates.

m/z 377	Control		WT		B2	
	exp	corr	exp	corr	exp	corr
M+0	0.82 ± 0.03	0.82 ± 0.02	0.07 ± 0.01	0.07 ± 0.01	0.05 ± 0.01	0.05 ± 0.01
M+1	0.15 ± 0.00	0.16 ± 0.00	0.02 ± 0.00	0.02 ± 0.00	0.01 ± 0.00	0.01 ± 0.00
M+2	0.02 ± 0.00	0.02 ± 0.00	0.00 ± 0.00	0.00 ± 0.00	0.01 ± 0.00	0.01 ± 0.00
M+3	0.00 ± 0.00	0.00 ± 0.00	0.00 ± 0.00	0.00 ± 0.00	0.02 ± 0.00	0.02 ± 0.00
M+4	0.00 ± 0.00	0.00 ± 0.00	0.00 ± 0.00	0.01 ± 0.00	0.03 ± 0.00	0.03 ± 0.00
M+5	0.00 ± 0.00	0.00 ± 0.00	0.01 ± 0.00	0.01 ± 0.00	0.05 ± 0.01	0.05 ± 0.01
M+6	0.00 ± 0.00	0.00 ± 0.00	0.02 ± 0.00	0.02 ± 0.00	0.07 ± 0.02	0.07 ± 0.02
M+7	0.00 ± 0.00	0.00 ± 0.00	0.03 ± 0.01	0.04 ± 0.01	0.08 ± 0.02	0.08 ± 0.02
M+8	0.00 ± 0.00	0.00 ± 0.00	0.07 ± 0.02	0.07 ± 0.02	0.12 ± 0.03	0.12 ± 0.03
M+9	0.00 ± 0.00	0.00 ± 0.00	0.07 ± 0.01	0.07 ± 0.01	0.10 ± 0.02	0.10 ± 0.02
M+10	0.00 ± 0.00	0.00 ± 0.00	0.09 ± 0.02	0.09 ± 0.02	0.11 ± 0.03	0.11 ± 0.03
M+11	0.00 ± 0.00	0.00 ± 0.00	0.15 ± 0.03	0.16 ± 0.03	0.12 ± 0.02	0.12 ± 0.02
M+12	0.00 ± 0.00	0.00 ± 0.00	0.17 ± 0.03	0.18 ± 0.03	0.11 ± 0.02	0.11 ± 0.02
M+13	0.00 ± 0.00	0.00 ± 0.00	0.25 ± 0.05	0.25 ± 0.05	0.11 ± 0.02	0.11 ± 0.02
M+14	0.00 ± 0.00	0.00 ± 0.00	0.03 ± 0.01	0.02 ± 0.01	0.01 ± 0.00	0.01 ± 0.00
M+15	0.00 ± 0.00	0.00 ± 0.00	0.01 ± 0.00	0.00 ± 0.00	0.00 ± 0.00	0.00 ± 0.00
M+16	0.00 ± 0.00	0.00 ± 0.00	0.00 ± 0.00	0.00 ± 0.00	0.00 ± 0.00	0.00 ± 0.00
M+17	0.00 ± 0.00	0.00 ± 0.00	0.00 ± 0.00	0.00 ± 0.00	0.00 ± 0.00	0.01 ± 0.00
SFL [%]	1.2 ± 0.10		59.1 ± 8.44		51.8 ± 9.10	
SFL$_{corr}$ [%]	**0.1 ± 0.01**		**59.3 ± 8.47**		**51.8 ± 9.10**	

Table 22: Relative mass isotopomer fractions of riboflavin (*m/z* 377) from culture supernatants of *A. gossypii* B2 grown on naturally labeled vegetable oil and 99 % [$^{13}C_2$] glycine or [^{13}C] formate. The control represents the labeling data for *A. gossypii* grown on fully naturally labeled medium. Time points indicate cultivation time after which the respective tracer was added. Data denote experimental LC/MS data (exp) and corrected labeling patterns (corr). Data were corrected for occurrence of natural isotopes. The mass isotopomer M+0 represents the relative amount of non-labeled, M+1 the amount of singly-labeled mass isotopomer fraction and so on. As higher mass isotopomers were below the detection limit, only the first eleven are listed. Data were obtained from three individual replicates.

m/z 377		Control	[$^{13}C_2$] Gly	[^{13}C] For	
Time [h]		0	0	0	48
M+0	exp	0.82 ± 0.03	0.16 ± 0.01	0.82 ± 0.04	0.67 ± 0.03
	corr	0.99 ± 0.03	0.21 ± 0.07	1.00 ± 0.05	0.82 ± 0.02
M+1	exp	0.15 ± 0.00	0.05 ± 0.00	0.16 ± 0.01	0.27 ± 0.02
	corr	0.01 ± 0.00	0.04 ± 0.01	0.00 ± 0.00	0.18 ± 0.00
M+2	exp	0.02 ± 0.00	0.60 ± 0.05	0.02 ± 0.00	0.05 ± 0.00
	corr	0.00 ± 0.00	0.74 ± 0.13	0.00 ± 0.00	0.00 ± 0.00
M+3	exp	0.00 ± 0.00	0.16 ± 0.01	0.00 ± 0.00	0.01 ± 0.00
	corr	0.00 ± 0.00	0.00 ± 0.00	0.00 ± 0.00	0.00 ± 0.00
M+4	exp	0.00 ± 0.00	0.03 ± 0.00	0.00 ± 0.00	0.00 ± 0.00
	corr	0.00 ± 0.00	0.00 ± 0.00	0.00 ± 0.00	0.00 ± 0.00
M+5	exp	0.00 ± 0.00	0.00 ± 0.00	0.00 ± 0.00	0.00 ± 0.00
	corr	0.00 ± 0.00	0.00 ± 0.00	0.00 ± 0.00	0.00 ± 0.00
M+6	exp	0.00 ± 0.00	0.00 ± 0.00	0.00 ± 0.00	0.00 ± 0.00
	corr	0.00 ± 0.00	0.00 ± 0.00	0.00 ± 0.00	0.00 ± 0.00
M+7	exp	0.00 ± 0.00	0.00 ± 0.00	0.00 ± 0.00	0.00 ± 0.00
	corr	0.00 ± 0.00	0.00 ± 0.00	0.00 ± 0.00	0.00 ± 0.00
M+8	exp	0.00 ± 0.00	0.00 ± 0.00	0.00 ± 0.00	0.00 ± 0.00
	corr	0.00 ± 0.00	0.00 ± 0.00	0.00 ± 0.00	0.00 ± 0.00
M+9	exp	0.00 ± 0.00	0.00 ± 0.00	0.00 ± 0.00	0.00 ± 0.00
	corr	0.00 ± 0.00	0.00 ± 0.00	0.00 ± 0.00	0.00 ± 0.00
M+10	exp	0.00 ± 0.00	0.00 ± 0.00	0.00 ± 0.00	0.00 ± 0.00
	corr	0.00 ± 0.00	0.00 ± 0.00	0.00 ± 0.00	0.00 ± 0.00
SFL	**[%]**	**0.0 ± 0.0**	**9.0 ± 0.8**	**0.0 ± 0.0**	**1.0 ± 0.1**

Table 23: Relative mass isotopomer fractions of riboflavin (*m/z* 377) from culture supernatants of *A. gossypii* B2 grown on naturally labeled vegetable oil and [3-^{13}C] serine, [^{13}C$_3$] serine, or [U^{13}C] yeast extract. Time points indicate cultivation time after which the respective tracer was added. Data denote experimental LC/MS data (exp) and corrected labeling patterns (corr). Data were corrected for occurrence of natural isotopes. The mass isotopomer M+0 represents the relative amount of non-labeled, M+1 the amount of singly-labeled mass isotopomer fraction and so on. As higher mass isotopomers were below the detection limit, only the first eleven are listed. Data were obtained from three individual replicates.

m/z 377		[3-^{13}C] Ser		[^{13}C$_3$] Ser		[U^{13}C] YE	
Time [h]		0	48	0	48	0-32	32
M+0	exp	0.80 ± 0.04	0.76 ± 0.04	0.73 ± 0.04	0.68 ± 0.03	0.63 ± 0.06	0.46 ± 0.05
	corr	0.98 ± 0.05	0.93 ± 0.05	0.88 ± 0.04	0.81 ± 0.07	0.65 ± 0.07	0.48 ± 0.05
M+1	exp	0.18 ± 0.01	0.21 ± 0.02	0.17 ± 0.01	0.18 ± 0.01	0.21 ± 0.02	0.21 ± 0.02
	corr	0.03 ± 0.00	0.07 ± 0.01	0.05 ± 0.00	0.08 ± 0.07	0.21 ± 0.02	0.21 ± 0.01
M+2	exp	0.03 ± 0.00	0.04 ± 0.00	0.08 ± 0.01	0.11 ± 0.01	0.07 ± 0.01	0.13 ± 0.01
	corr	0.00 ± 0.00	0.00 ± 0.00	0.07 ± 0.01	0.09 ± 0.01	0.06 ± 0.01	0.13 ± 0.01
M+3	exp	0.00 ± 0.00	0.00 ± 0.00	0.02 ± 0.00	0.03 ± 0.00	0.03 ± 0.00	0.06 ± 0.01
	corr	0.00 ± 0.00	0.00 ± 0.00	0.00 ± 0.00	0.02 ± 0.00	0.02 ± 0.00	0.05 ± 0.00
M+4	exp	0.00 ± 0.00	0.00 ± 0.00	0.00 ± 0.00	0.01 ± 0.00	0.03 ± 0.00	0.06 ± 0.01
	corr	0.00 ± 0.00	0.00 ± 0.00	0.00 ± 0.00	0.00 ± 0.00	0.03 ± 0.00	0.06 ± 0.01
M+5	exp	0.00 ± 0.00	0.00 ± 0.00	0.00 ± 0.00	0.00 ± 0.00	0.01 ± 0.00	0.03 ± 0.00
	corr	0.00 ± 0.00	0.00 ± 0.00	0.00 ± 0.00	0.00 ± 0.00	0.01 ± 0.00	0.03 ± 0.00
M+6	exp	0.00 ± 0.00	0.00 ± 0.00	0.00 ± 0.00	0.00 ± 0.00	0.00 ± 0.00	0.01 ± 0.00
	corr	0.00 ± 0.00	0.00 ± 0.00	0.00 ± 0.00	0.00 ± 0.00	0.00 ± 0.00	0.01 ± 0.00
M+7	exp	0.00 ± 0.00	0.00 ± 0.00	0.00 ± 0.00	0.00 ± 0.00	0.00 ± 0.00	0.01 ± 0.00
	corr	0.00 ± 0.00	0.00 ± 0.00	0.00 ± 0.00	0.00 ± 0.00	0.00 ± 0.00	0.01 ± 0.00
M+8	exp	0.00 ± 0.00	0.00 ± 0.00	0.00 ± 0.00	0.00 ± 0.00	0.00 ± 0.00	0.01 ± 0.00
	corr	0.00 ± 0.00	0.00 ± 0.00	0.00 ± 0.00	0.00 ± 0.00	0.00 ± 0.00	0.01 ± 0.00
M+9	exp	0.00 ± 0.00	0.00 ± 0.00	0.00 ± 0.00	0.00 ± 0.00	0.01 ± 0.00	0.01 ± 0.00
	corr	0.00 ± 0.00	0.00 ± 0.00	0.00 ± 0.00	0.00 ± 0.00	0.01 ± 0.00	0.01 ± 0.00
M+10	exp	0.00 ± 0.00	0.00 ± 0.00	0.00 ± 0.00	0.00 ± 0.00	0.00 ± 0.00	0.00 ± 0.00
	corr	0.00 ± 0.00	0.00 ± 0.00	0.00 ± 0.00	0.00 ± 0.00	0.00 ± 0.00	0.00 ± 0.00
SFL	[%]	0.2 ± 0.1	0.4 ± 0.0	1.2 ± 0.1	1.9 ± 0.2	4.5 ± 0.6	8.0 ± 1.1

6.8 Data from NMR analyses

Table 24: Relative ^{13}C enrichment of all seventeen carbon atoms of riboflavin produced by *A. gossypii* B2 from different ^{13}C-labeled tracer substrates on complex medium with rapeseed oil. Riboflavin was recovered from the culture broth after 144 h. The labeling was analyzed by ^{13}C NMR. The time refers to the time point of respective tracer addition. For, formate; Glu, glutamate; Gly, Glycine; Ser, serine; YE, yeast extract. Data denote mean values for three independent replicates with a mean standard deviation of 5 %.

C-atom	Chemical shift [ppm]	Nat. lab. precursors	[^{13}C] For		[^{13}C$_2$] Gly	[3-^{13}C] Ser	[U^{13}C] YE	[^{13}C$_4$] Glu
		0 h	0 h	48 h	0 h	48 h	32 h	0 h
2	155.5	1.1	4.7	12.3	1.1	5.9	15.4	2.1
4	159.9	1.1	1.1	1.1	1.0	1.0	11.8	2.8
4a	136.8	1.1	1.3	0.9	71.2	0.9	18.4	1.1
5a	134	1.1	1.1	0.9	1.4	1.0	4.8	2.9
6	130	1.1	1.1	1.1	1.0	1.1	5.0	3.3
7	137.1	1.1	0.9	0.7	1.3	0.8	3.7	2.8
7α	18.8	1.3	1.1	1.2	1.1	1.1	3.8	2.7
8	146	1.1	1.0	0.9	1.3	0.9	4.6	2.9
8α	20.8	1.3	1.1	1.2	1.1	1.1	4.7	3.6
9	117.4	1.0	1.5	1.0	0.9	1.2	3.9	1.8
9a	132.1	1.1	1.1	0.9	1.6	0.9	3.7	1.4
10a	150.8	1.1	1.0	0.8	70.7	0.9	18.5	1.0
1'	47.3	1.2	1.1	1.0	1.0	1.0	9.9	3.9
2'	68.8	1.1	1.1	1.0	1.2	1.0	8.8	5.0
3'	73.6	1.2	1.1	0.9	1.5	0.9	7.9	3.9
4'	72.8	1.3	1.1	1.0	0.8	0.9	8.3	4.8
5'	63.4	1.0	1.1	1.1	0.8	1.1	7.5	2.7

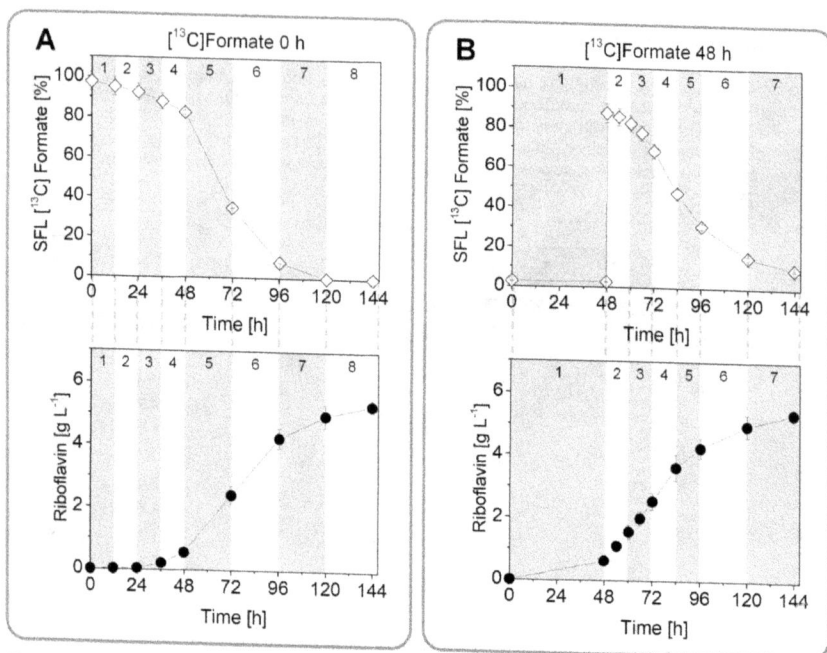

Figure 42: Change in summed fractional labeling (SFL) for formate added at 0 h (A) and formate added at 48 h (B) in the culture supernatant of *A. gossypii* B2 grown on complex medium, rapeseed oil, and [13C] formate. Labeling data for formate were measured via 1H NMR. Riboflavin data were interpolated for certain points in time. Numbers indicate single intervals, for which the maximum 13C enrichment of riboflavin at carbon atom C_2 was calculated (Chapters 3.5.2 and 4.6.1). Data were obtained from three individual replicates.

6.9 Data from ^{13}C formate flux simulations

Table 25: Input model for OpenFLUX for the transmembrane formate flux.

rxnID	rxnEQ	cTrans	rates	type	basis
R01	FOR_FEED = FOR_EXTR	a = a		F	100.0
R02	FOR_EXTR = FOR_INTR	a = a		FR	
R03	FOR_INTR = FOR_EXTR	a = a		R	X
R04	FOR_P5P = FOR_INTR	a = a		F	X
R05	FOR_INTR = FOR_RF			B	
R06	FOR_EXTR = FOR_EX			B	
excludedMetabolites					
FOR_FEED					
FOR_P5P					
FOR_RF					
FOR_EX					
simulatedMDVs					
FOR_EXTR#1					
FOR_INTR#1					
inputSubstrates					
FOR_P5P					
FOR_FEED					

RxnID, reaction identification; cTrans, carbon transition; rates, known reaction rates, type, reaction type (F = irreversible, FR = forward reaction of a reversible reaction, R = backward reaction, B = excluded from the isotopomer balance); basis, invariant flux basis; excludedMetabolites, metabolites excluded from the balance model; simulatedMDVs, simulated mass isotopomer distribution vectors; inputSubstrates, compounds that are used as substrates.

Table 26: Relative mass isotopomer fractions of formate from culture supernatants and flux simulation of *A. gossypii* B2 grown on vegetable oil and 99 % [^{13}C] formate added at 0 h. Data denote experimental ^{1}H NMR data (exp) and simulated labeling patterns (sim) corresponding to the optimal flux fit. The mass isotopomer M+0 represents the relative amount of non-labeled, M+1 the amount of singly-labeled mass isotopomer fraction. The time intervals t1, t2 etc. correspond to the intervals shown in Figure 34. Data denote mean values from independent replicates with a mean standard deviation of 5 %.

			t1	t2	t3	t4	t5
For$_{EXTR}$	M+0	exp	0.037	0.084	0.145	0.410	0.786
		sim	0.037	0.084	0.145	0.410	0.786
	M+1	exp	0.963	0.916	0.855	0.590	0.214
		sim	0.963	0.916	0.855	0.590	0.214
FOR$_{INTR}$	M+0	exp	0.037	0.084	0.436	0.855	0.976
		sim	0.037	0.085	0.441	0.858	0.976
	M+1	exp	0.963	0.916	0.564	0.145	0.024
		sim	0.963	0.915	0.559	0.142	0.024

6.10 Determination of measured carbon fluxes and carbon balances for carbon flux calculations during the riboflavin biosynthetic phase of *A. gossypii*

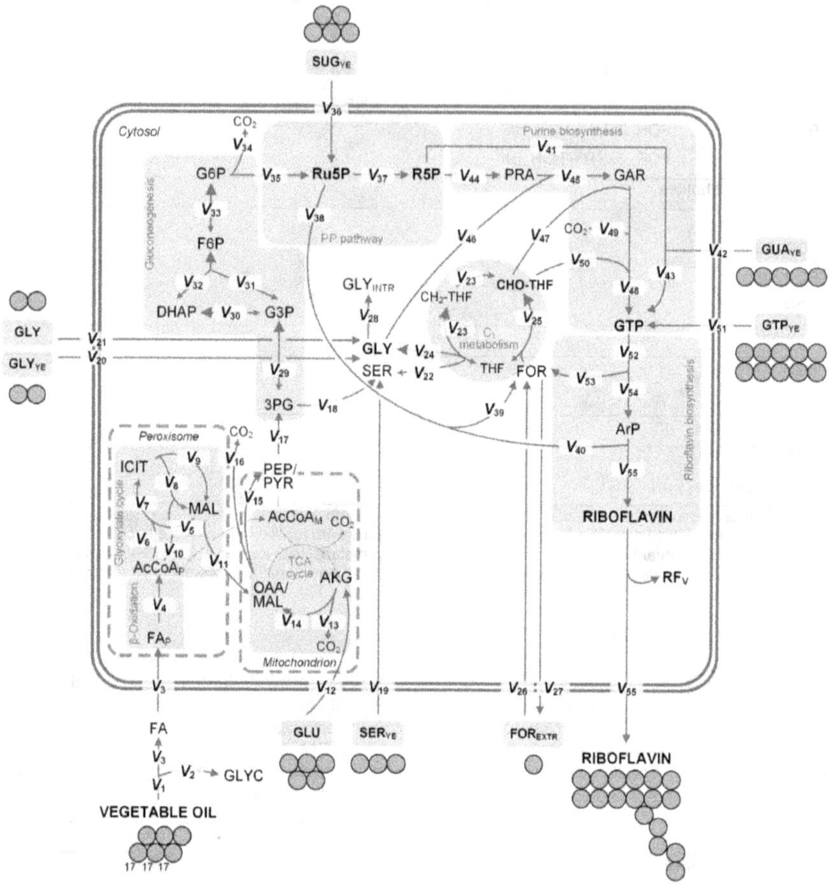

Figure 43: Metabolic network of riboflavin production in *A. gossypii*, including extracellular reactions, reactions between intermediary metabolite pools, and riboflavin biosynthetic reactions. The reaction numbers refer to the formulation of metabolite balances (see below). The direction of net reactions is indicated by size of arrow head. For abbreviations see Figure 37.

During riboflavin biosynthesis on vegetable oil (Figure 43), the fluxes to be determined were calculated and expressed as fluxes of single carbon atoms. This facilitated flux calculations, because every carbon atom of riboflavin had a unique labeling fingerprint obtained from the four tracer studies with [$^{13}C_2$] glycine, [^{13}C] formate, [$^{13}C_5$] glutamate, and [$U^{13}C$] yeast extract. Since a single carbon atom of riboflavin could originate from a number of different medium components as well as intracellular metabolites, single carbon atom balancing met the challenges of complex data handling the best. Therefore, for every carbon atom in riboflavin, the metabolic precursor carbon atom and other donor atoms were considered (Figure 45, Figure 46, Figure 47) and the flux (v) into the respective carbon atom was set to 1 and formulated as follows:

$$v_i = \sum_{j=1}^{5} v_{i,j} = 1 \qquad \text{(Eq. 44)}$$

$$v_{i,j} = x_{i,j} \qquad \text{(Eq. 45)}$$

$$v_{i,Gly} = x_{i,Gly} \qquad \text{(Eq. 46)}$$

$$v_{i,For} = x_{i,For} \qquad \text{(Eq. 47)}$$

$$v_{i,Glu} = x_{i,Glu} \qquad \text{(Eq. 48)}$$

$$v_{i,YE} = x_{i,YE} \qquad \text{(Eq. 49)}$$

$$v_{i,Oil} = 1 - v_{i,Gly} - v_{i,For} - v_{i,Glu} - v_{i,YE} \qquad \text{(Eq. 50)}$$

with

v_i	carbon atom of riboflavin with number i
j	index specifying the ^{13}C tracer used in each of the four parallel isotope experiments (1: glycine; 2: formate; 3: glutamate; 4: yeast extract (YE); 5: oil)
x_i	specific ^{13}C enrichment at carbon atom i of riboflavin from the corresponding ^{13}C tracer j

Note that oil was not used as ^{13}C tracer, but the enrichment (x_{Oil}) was a result of the combined enrichment of all ^{13}C tracers used (formate, glycine, glutamate, and yeast extract). The Equations (44) to (50) could then be applied in detail to every carbon atom in riboflavin. Carbon atom C_4 of riboflavin will be discussed exemplary in greater detail below, followed by the other carbon atoms of riboflavin, however, to a less extensive extent.

Considerations regarding carbon atom **C₄**:

Equation (51) was specified for carbon atom C₄, based on Equation (44):

$$v_4 = \sum_{j=1}^{5} v_{4,j} = v_{4,Gly} + v_{4,For} + v_{4,Glu} + v_{4,YE} + v_{4,Oil} = 1 \qquad \text{(Eq. 51)}$$

The carbon origin of C₄ of riboflavin is depicted in Figure 44. Since its biosynthetic *de novo* origin is carbon dioxide, the oil and the glutamate fraction of Equation (51) are defined as fraction derived from carbon dioxide instead, which is why the combined flux equals flux V_{49} of the network (Equation 52) (Figure 43), i.e. the uptake of single carbon atoms from oil. Carbon dioxide is produced from rapeseed oil (naturally labeled) and glutamate (^{13}C-labeled) through decarboxylation reactions in the TCA cycle as well as at the pyruvate node. Note that the equality in Equation (52) is straightforward, as only single carbon atoms are considered.

$$v_{4,CO_2} = v_{4,Oil} + v_{4,Glu} = V_{49} \qquad \text{(Eq. 52)}$$

For the carbon flux from the yeast extract to C₄, two potential building blocks were considered: GTP (or ATP, however, since this study could not distinguish between the two, only GTP is used) or guanine (or adenine). GTP as well as guanine are part of the yeast extract and are incorporated into riboflavin almost as whole molecule only releasing one carbon atom (Figure 32). Therefore, the flux from the yeast extract could be formulated as follows:

$$v_{4,YE} = v_{4,YE_{GTP}} + v_{4,YE_{GUA}} \qquad \text{(Eq. 53)}$$

with

YE_{GTP} GTP (or ATP) originating from ^{13}C-labeled yeast extract

YE_{GUA} guanine (or adenine) originating from ^{13}C-labeled yeast extract

GTP shares nine of its ten carbon atoms with riboflavin. The remaining eight carbon atoms comprise the xylene ring and are exclusive carbon atoms for the vitamin. Thus, the carbon flux from GTP from the yeast extract can be expressed as the difference between ^{13}C labeling from [U^{13}C] yeast extract of carbon atoms of the ribityl side chain (C₁' to C₅') and carbon atoms of the xylene ring (C₅ₐ to C₉ₐ). In order to simplify the complexity, mean values derived from the structural subunits were used ($\overline{v_{1'-5',YE}}$ and $\overline{v_{5a-9a,YE}}$, respectively).

$$v_{4,YE} = \left(\overline{v_{1'-5',YE}} - \overline{v_{5a-9a,YE}} \right) + v_{4,YE_{GUA}} \qquad \text{(Eq. 54)}$$

Rearrangement of Equation (54) leads to the carbon flux from yeast extract-based guanine to carbon atom C_4:

$$v_{4,YE_{GUA}} = v_{4,YE} - \overline{v_{1'-5',YE}} + \overline{v_{5a-9a,YE}} \qquad \text{(Eq. 55)}$$

Since guanine is incorporated into riboflavin as a whole, it can be assumed that the carbon flux from yeast extract-derived guanine into C_4 also equals the carbon flux into the other three carbon atoms of riboflavin that originate from guanine (C_2, C_{4a}, C_{10a}). This correlation is expressed by Equation (56).

$$v_{4,YE_{GUA}} = v_{2,YE_{GUA}} = v_{4a,YE_{GUA}} = v_{10a,YE_{GUA}} \qquad \text{(Eq. 56)}$$

In a likewise manner, this can be applied to the carbon flux from yeast extract-based GTP to carbon atom C_4:

$$v_{4,YE_{GTP}} = v_{i,YE_{GTP}} \qquad \text{with } i = 2, 4a, 10a, 1' - 5' \qquad \text{(Eq. 57)}$$

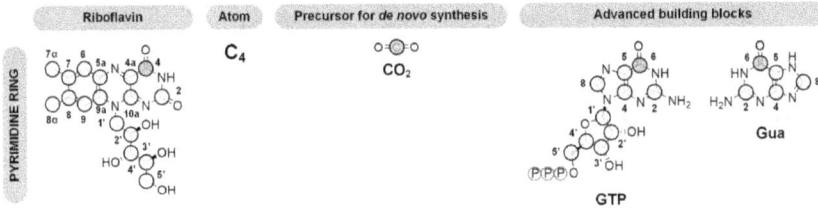

Figure 44: Metabolic origin of carbon atom C_4 of the pyrimidine ring in riboflavin. Grey circles denote the carbon atom of interest and its origin in the metabolic precursor or donor to the precursor. The numbering is specific for the molecule, therefore, e.g. carbon atom 4 of riboflavin does not equal carbon atom 4 of GTP. GTP, guanosine triphosphate; Gua, guanine. This is an excerpt of Figure 45.

Considerations regarding reactions involved in the formation of carbon atom C_2:

$$v_2 = \sum_{j=1}^{5} v_{2,j} = 1 \qquad \text{(Eq. 58)}$$

$$v_{2,YE} = v_{2,YE_{GTP}} + v_{2,YE_{GUA}} + v_{2,YE_{SER}} \qquad \text{(Eq. 59)}$$

$$v_{2,YE_{SER}} = v_{2,YE} - v_{2,YE_{GTP}} - v_{2,YE_{GUA}} = v_{4a,YE_{SER}} = v_{10a,YE_{SER}} \qquad \text{(Eq. 60)}$$

$$v_{2,Oil} = v_{2,Oil_{Ru5P}} + v_{2,Oil_{SER}} \qquad \text{(Eq. 61)}$$

$$v_{2,Oil_{SER}} = v_{4a,Oil} \qquad \text{(Eq. 62)}$$

$$v_{2,Oil_{Ru5P}} = v_{2,Oil} - v_{4a,Oil} \qquad \text{(Eq. 63)}$$

with

Oil_{Ru5P}	ribulose 5-phosphate originating from vegetable oil
Oil_{SER}	serine originating from vegetable oil, i.e. 3PG
YE_{SER}	serine originating from ^{13}C-labeled yeast extract

Considerations regarding reactions involved in the formation of carbon atom C_{4a}:

$$v_{4a} = \sum_{j=1}^{5} v_{4a,j} = 1 \qquad \text{(Eq. 64)}$$

$$v_{4a,YE} = v_{4a,YE_{GTP}} + v_{4a,YE_{GUA}} + v_{4a,YE_{SER}} + v_{4a,YE_{Gly}} \qquad \text{(Eq. 65)}$$

$$v_{4a,Oil} = v_{4a,Oil_{Ser}} \qquad \text{(Eq. 66)}$$

with

YE_{GLY}	glycine originating from ^{13}C-labeled yeast extract

Since the considerations for the formation of carbon atom C_{10a} equal the ones for carbon atom C_{4a}, they will not be described.

Considerations regarding reactions involved in the formation of carbon atom C_{5a}:

$$v_{5a} = \sum_{j=1}^{5} v_{5a,j} = 1 \qquad \text{(Eq. 67)}$$

$$v_{5a,Ru5P} = v_{5a,Oil} + v_{5a,Glu} + v_{5a,YE} = 1 \qquad \text{(Eq. 68)}$$

$$v_{5a,de\,novo} = v_{5a,Oil} + v_{5a,Glu} \qquad \text{(Eq. 69)}$$

$$v_{5a,de\,novo} = \frac{1}{2}v_{5a,OAA} = v_{5a,G6P} \qquad \text{(Eq. 70)}$$

with

$V_{5a,de\ novo}$	C_{5a} atom originating from *de novo* biosynthesis through glyoxylate or TCA cycle and gluconeogenesis
$V_{5a,OAA}$	C_{5a} atom originating from the C_2 atom of oxaloacetate
$V_{5a,G6P}$	C_{5a} atom originating from the C_2 atom of glucose 6-phosphate

Since the considerations for the formation of carbon atom C_5 through C_{9a} occurred in a likewise manner as the ones for carbon atom C_{5a}, they will not be described.

Considerations regarding reactions involved in the formation of carbon atom $\mathbf{C_{1'}}$:

$$v_{1'} = \sum_{j=1}^{5} v_{1',j} = 1 \qquad \text{(Eq. 71)}$$

$$v_{1',YE} = v_{1',YE_{SUGAR}} + v_{1',YE_{GTP}} \qquad \text{(Eq. 72)}$$

$$v_{1',Ru5P} = v_{1',Oil} + v_{1',YE_{SUGAR}} + v_{1',Glu} \qquad \text{(Eq. 73)}$$

$$v_{1',de\ novo} = v_{1',Oil} + v_{1',Glu} \qquad \text{(Eq. 74)}$$

$$v_{1',de\ novo} = \frac{1}{2}v_{1',OAA} = v_{1',G6P} \qquad \text{(Eq. 75)}$$

with

YE_{SUGAR}	Sugar originating from ^{13}C-labeled yeast extract
$V_{1',de\ novo}$	$C_{1'}$ atom originating from *de novo* biosynthesis through glyoxylate or TCA cycle and gluconeogenesis
$V_{1',OAA}$	$C_{1'}$ atom originating from the C_1 atom of oxaloacetate
$V_{1',G6P}$	$C_{1'}$ atom originating from the C_1 atom of glucose 6-phosphate

Since the considerations for the formation of carbon atom $C_{2'}$ through $C_{5'}$ equal the ones for carbon atom $C_{1'}$, they will not be described.

Based on the carbon atom fluxes, the calculation of fluxes between metabolites (designated by a capital *V*), i.e. a sum of several carbon atoms, could then be formulated for the three major structural units that comprise the vitamin (Figure 45, Figure 46, Figure 47). Note that riboflavin contains 17 carbon atoms in total, the xylene ring eight carbon atoms, the ribityl side chain five carbon atoms, and the pyrimidine ring four carbon atoms:

$$V_{55} = V_{RF} = \sum_{i=1}^{17} v_i = \sum_{i=1}^{17} \sum_{j=1}^{5} v_{i,j} = 17 \qquad \text{(Eq. 76)}$$

$$V_{40} = V_{Xylene} = \sum_{i=5a}^{9a} v_i = \sum_{i=5a}^{9a} \sum_{j=1}^{5} v_{i,j} = 8 \qquad \text{(Eq. 77)}$$

$$V_{Ribityl} = \sum_{i=1\prime}^{5\prime} v_i = \sum_{i=1\prime}^{5\prime} \sum_{j=1}^{5} v_{i,j} = 5 \qquad \text{(Eq. 78)}$$

$$V_{Pyrimidine} = v_2 + v_4 + v_{4a} + v_{10a} = 4 \qquad \text{(Eq. 79)}$$

$$V_{54} = V_{Ribityl} + V_{Pyrimidine} \qquad \text{(Eq. 80)}$$

These sums of carbon atoms provide three out of 55 pieces of information required for a fully determined metabolic network that consisted of a total of 55 fluxes (V_{55}, V_{40}, V_{54}) (Figure 43). Another piece of information was defined in Equation (52) (V_{49}). The labeling data for all seventeen carbon atoms of riboflavin from four different labeling experiments (Table 14 and appendix Table 24) and underlying pathway stoichiometry allowed the calculation of 28 fluxes. The calculation of those fluxes is given in Equations (81) to (107) (V_2, V_8, V_9, V_{12}, V_{13}, V_{16}, V_{18}, V_{19}, V_{20}, V_{21}, V_{22}, V_{23}, V_{24}, V_{25}, V_{26}, V_{34}, V_{36}, V_{37}, V_{38}, V_{39}, V_{41}, V_{42}, V_{43}, V_{46}, V_{47}, V_{50}, V_{51}, V_{53}). It should be noted that most of the numbers are derived from carbon atom stoichiometry (e.g. 52_{FA} designates that FA contains 52 carbon atoms, if it is defined as three FA chains of average length of 17.3 carbon atoms). In those cases, the origin of those carbon atoms is specified by the index (for abbreviations refer to Chapter 6.1). Some carbon fluxes are presented as mean carbon flux from a structural group, e.g. xylene ring ($\overline{v_{5a-9a,Glu}}$ and $\overline{v_{5a-9a,YE}}$) or two-carbon unit of riboflavin ($\overline{v_{4a;10a,Gly}}$ and $\overline{v_{4a;10a,YE_{Gly}}}$). This was done in order to simplify the calculations and does not affect their outcome. The carbon fluxes v_{GUA_8} in Equation (102) and v_{GTP_8} in Equation (106) refer to the flux of the carbon atom number 8 in guanine or GTP (Figure 44), respectively. Their values equal the ones in Equation (56) and Equation (57), respectively.

$$V_2 = V_3 \cdot \frac{3_{GLYC}}{52_{FA}} \qquad \text{(Eq. 81)}$$

$$V_8 = \frac{V_{11}}{2_{GLYOX}} \qquad \text{(Eq. 82)}$$

$$V_9 = V_{11} \qquad \text{(Eq. 83)}$$

$$V_{12} = V_{15} \cdot \frac{\overline{v_{5a-9a,Glu}}}{1 - \overline{v_{5a-9a,YE}}} \cdot \frac{5_{GLU}}{4_{OAA}} \qquad \text{(Eq. 84)}$$

$$V_{13} = \frac{V_{12}}{5_{GLU}} \qquad \text{(Eq. 85)}$$

$$V_{16} = \frac{V_{15}}{4_{OAA}} \tag{Eq. 86}$$

$$V_{18} = V_{22} \cdot \frac{v_{2,Oil_{SER}}}{v_{2,YE_{SER}} + v_{2,Oil_{SER}}} \tag{Eq. 87}$$

$$V_{19} = V_{22} \cdot \frac{v_{2,YE_{SER}}}{v_{2,YE_{SER}} + v_{2,Oil_{SER}}} \tag{Eq. 88}$$

$$V_{20} = (V_{28} + V_{46}) \cdot \frac{\overline{v}_{4a;10a,Gly}}{1 - \overline{v}_{4a;10a,YE_{GTP}} - \overline{v}_{4a;10a,YE_{GUA}}} \tag{Eq. 89}$$

$$V_{21} = (V_{28} + V_{46}) \cdot \frac{\overline{v}_{4a;10a,YE_{Gly}}}{1 - \overline{v}_{4a;10a,YE_{GTP}} - \overline{v}_{4a;10a,YE_{GUA}}} \tag{Eq. 90}$$

$$V_{22} = 3_{SER} \cdot V_{23} \tag{Eq. 91}$$

$$V_{23} = 2 \cdot \left(v_{2,YE_{SER}} + v_{2,Oil_{SER}}\right) \tag{Eq. 92}$$

$$V_{24} = 2_{GLY} \cdot V_{23} \tag{Eq. 93}$$

$$V_{25} = 2 \cdot \left(v_{2,Oil_{Ru5P}} + v_{2,For}\right) \tag{Eq. 94}$$

$$V_{26} = V_{39} \cdot \frac{v_{2,For}}{v_{2,Oil_{Ru5P}}} \tag{Eq. 95}$$

$$V_{34} = \frac{V_{33}}{6_{G6P}} \tag{Eq. 96}$$

$$V_{36} = \sum_{i=1'}^{5'} v_{i,YE_{SUGAR}} + \sum_{i=5a}^{9a} v_{i,YE} \tag{Eq. 97}$$

$$V_{37} = \sum_{i=1'}^{5'} v_{i,Ru5P} \tag{Eq. 98}$$

$$V_{38} = V_{39} + V_{40} \tag{Eq. 99}$$

$$V_{39} = \frac{V_{40}}{4_{DHBP}} \tag{Eq. 100}$$

$$V_{41} = V_{42} \tag{Eq. 101}$$

$$V_{42} = v_{2,YE_{GUA}} + v_{4,YE_{GUA}} + v_{4a,YE_{GUA}} + v_{10a,YE_{GUA}} + v_{1',YE_{GUA}} \tag{Eq. 102}$$
$$+ v_{2',YE_{GUA}} + v_{3',YE_{GUA}} + v_{4',YE_{GUA}} + v_{5',YE_{GUA}} + v_{GUA_8}$$

$$V_{43} = V_{41} + V_{42} \tag{Eq. 103}$$

$$V_{46} = \left(v_{4a} - v_{4a,YE_{GUA}} - v_{4a,YE_{GTP}}\right) \tag{Eq. 104}$$
$$+ \left(v_{10a} - v_{10a,YE_{GUA}} - v_{10a,YE_{GTP}}\right)$$

$$V_{47} = V_{50} = v_2 - v_{2,YE_{GUA}} - v_{2,YE_{GTP}} \tag{Eq. 105}$$

$$V_{51} = v_{2,YE_{GTP}} + v_{4,YE_{GTP}} + v_{4a,YE_{GTP}} + v_{10a,YE_{GTP}} + v_{1',YE_{GTP}} \tag{Eq. 106}$$
$$+ v_{2',YE_{GTP}} + v_{3',YE_{GTP}} + v_{4',YE_{GTP}} + v_{5',YE_{GTP}} + v_{GTP_8}$$

$$V_{53} = \frac{V_{52}}{10_{GTP}} \tag{Eq. 107}$$

Together with the four fluxes determined above (V_{40}, V_{49}, V_{54}, V_{55}) this rendered 32 fluxes. Thus, at least 23 pieces of information were still required to obtain a fully determined network. Therefore, following balances were formulated for the metabolic network depicted in Figure 43, which represents riboflavin production on vegetable oil and yeast extract. Metabolite balances were expressed using the numbering of the reactions as presented in Figure 43.

FA	$0 = V_1 - V_2 - V_3$	(Eq. 108)
FA$_P$	$0 = V_3 - V_4$	(Eq. 109)
AcCoA$_P$	$0 = V_4 - V_6 - V_{10}$	(Eq. 110)
ICIT	$0 = V_7 - V_8 - V_9$	(Eq. 111)
MAL	$0 = V_9 + V_{10} + V_8 - V_5 - V_{11}$	(Eq. 112)
OAA/MAL	$0 = V_{11} + V_{14} - V_{15} - V_{16}$	(Eq. 113)
AKG	$0 = V_{12} - V_{13} - V_{14}$	(Eq. 114)
PEP/PYR	$0 = V_{15} - V_{17}$	(Eq. 115)
3PG	$0 = V_{17} - V_{18} - V_{29}$	(Eq. 116)
G3P	$0 = V_{29} - V_{30} - V_{31}$	(Eq. 117)
DHAP	$0 = V_{30} - V_{32}$	(Eq. 118)
F6P	$0 = V_{31} + V_{32} - V_{33}$	(Eq. 119)
G6P	$0 = V_{33} - V_{34} - V_{35}$	(Eq. 120)
Ru5P	$0 = V_{35} + V_{36} - V_{37} - V_{38}$	(Eq. 121)
R5P	$0 = V_{37} - V_{41} - V_{44}$	(Eq. 122)
PRA	$0 = V_{44} + V_{46} - V_{45}$	(Eq. 123)
GAR	$0 = V_{45} + V_{47} + V_{49} + V_{50} - V_{48}$	(Eq. 124)
GTP	$0 = V_{43} + V_{48} + V_{51} - V_{52}$	(Eq. 125)
ArP	$0 = V_{40} + V_{54} - V_{55}$	(Eq. 126)
GLY	$0 = V_{20} + V_{21} + V_{24} - V_{28} - V_{46}$	(Eq. 127)
SER	$0 = V_{18} + V_{19} - V_{22}$	(Eq. 128)
FOR	$0 = V_{26} + V_{39} + V_{53} - V_{25} - V_{27}$	(Eq. 129)
CHO-THF	$0 = V_{23} + V_{25} - V_{47} - V_{50}$	(Eq. 130)

The rank of the stoichiometric matrix formulated for Equations (108) to (130) was 23. This demonstrated that the 23 metabolite balances were linearly independent. Since the complete network comprised a total of 55 fluxes the combined 55 pieces of information obtained through metabolite balancing and labeling information rendered a fully determined network for flux calculations.

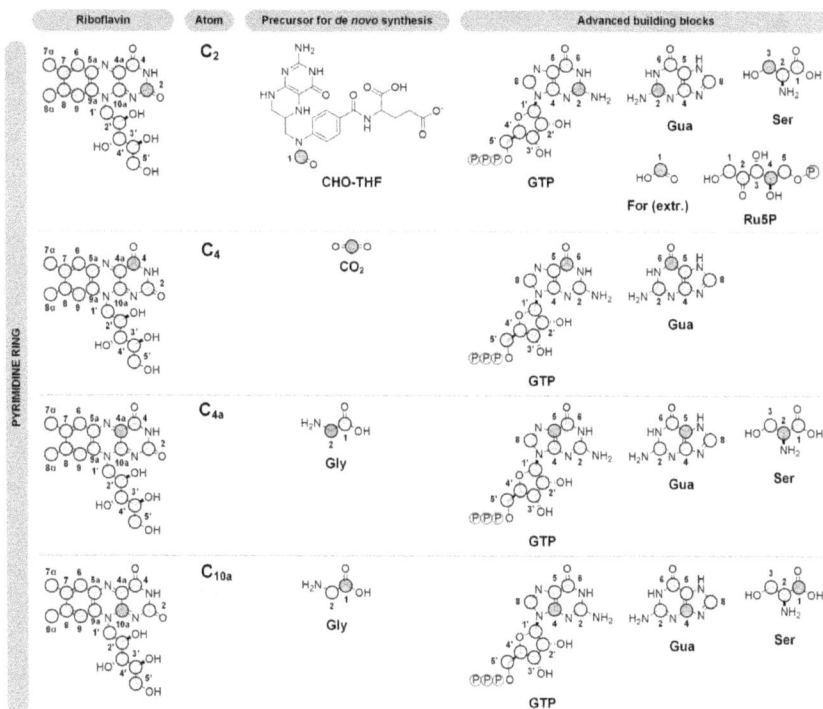

Figure 45: Metabolic origin of carbon atoms of the pyrimidine ring in riboflavin. Grey circles denote the carbon atom of interest and its origin in the metabolic precursor or donor to the precursor. The numbering is specific for the molecule, therefore, e.g. carbon atom 2 of riboflavin does not equal carbon atom 2 of GTP. CHO-THF, 10-formyltetrahydrofolate; For (extr.), formate (extracellular); Gly, glycine; GTP, guanosine triphosphate; Gua, guanine; Ru5P, ribulose 5-phosphate; Ser, serine.

145

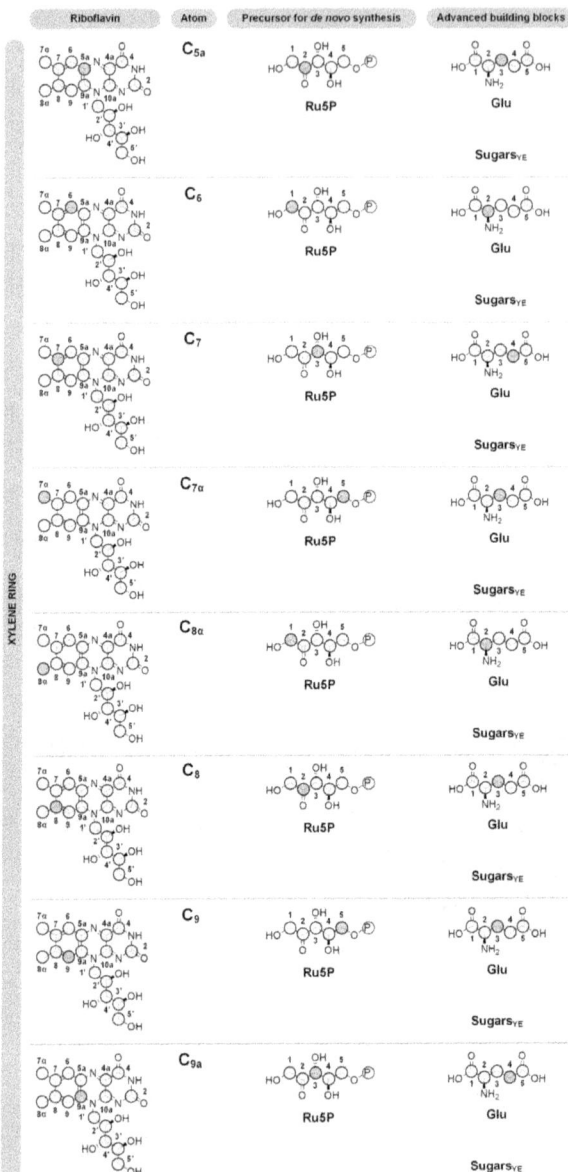

Figure 46: Metabolic origin of carbon atoms of the xylene ring in riboflavin. Grey circles denote the carbon atom of interest and its origin in the metabolic precursor or donor to the precursor. The numbering is specific for the molecule, therefore, e.g. carbon atom 2 of riboflavin does not equal carbon atom 2 of Ru5P. Glu, glutamate; Ru5P, ribulose 5-phosphate; Ser, serine; Sugars$_{YE}$, sugars derived from yeast extract.

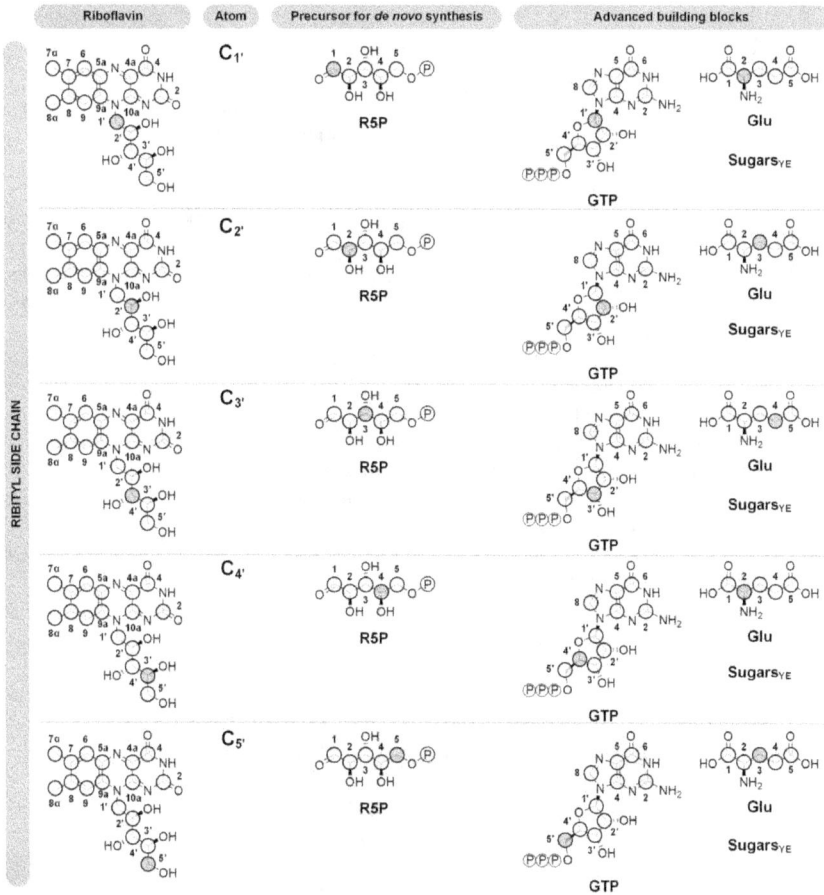

Figure 47: Metabolic origin of carbon atoms of the ribityl side chain of riboflavin. Grey circles denote the carbon atom of interest and its origin in the metabolic precursor or donor to the precursor. The numbering is specific for the molecule, therefore, e.g. carbon atom 2 of riboflavin does not equal carbon atom 2 of GTP. Glu, glutamate; GTP, guanosine triphosphate; R5P, ribose 5-phosphate; Sugars$_{YE}$, sugars derived from yeast extract.

7 REFERENCES

Abbas, C. A., Sibirny, A. A., 2011. Genetic control of biosynthesis and transport of riboflavin and flavin nucleotides and construction of robust biotechnological producers. Microbiol. Mol. Biol. Rev. 75, 321-360.

Adler, P., Bolten, C. J., Dohnt, K., Hansen, C. E., Wittmann, C., 2013. Core fluxome and metafluxome of lactic acid bacteria under simulated cocoa pulp fermentation conditions. Appl. Environ. Microbiol. 79, 5670-5681.

Adler, P., Frey, L. J., Berger, A., Bolten, C. J., Hansen, C. E., Wittmann, C., 2014. The key to acetate: metabolic fluxes of acetic acid bacteria under cocoa pulp fermentation-simulating conditions. Appl. Environ. Microbiol. 80, 4702-4716.

Aguiar, T. Q., Dinis, C., Magalhaes, F., Oliveira, C., Wiebe, M. G., Penttilä, M., Domingues, L., 2014a. Molecular and functional characterization of an invertase secreted by Ashbya gossypii. Mol. Biotechnol. 56, 524-534.

Aguiar, T. Q., Ribeiro, O., Arvas, M., Wiebe, M. G., Penttilä, M., Domingues, L., 2014b. Investigation of protein secretion and secretion stress in Ashbya gossypii. BMC Genomics. 15.

Aguiar, T. Q., Silva, R., Domingues, L., 2017. New biotechnological applications for Ashbya gossypii: Challenges and perspectives. Bioengineered. 8, 309-315.

Alder, L., Greulich, K., Kempe, G., Vieth, B., 2006. Residue analysis of 500 high priority pesticides: better by GC-MS or LC-MS/MS? Mass Spectrom. Rev. 25, 838-65.

Antoniewicz, M. R., Kelleher, J. K., Stephanopoulos, G., 2007. Elementary metabolite units (EMU): a novel framework for modeling isotopic distributions. Metab. Eng. 9, 68-86.

Ashby, S. F., Nowell, W., 1926. The fungi of stigmatomycosis. Ann. Bot. 40, 69-84.

Babel, W., Müller, R. H., Markuske, K. D., 1983. Improvement of growth yield of yeast on glucose to the maximum by using an additional energy source. Arch. Microbiol. 136, 203-208.

Bacher, A., Eberhardt, S., Fischer, M., Kis, K., Richter, G., 2000. Biosynthesis of vitamin B2 (riboflavin). Annu. Rev. Nutr. 20, 153-167.

Bacher, A., Levan, Q., Buhler, M., Keller, P. J., Eimicke, V., Floss, H. G., 1982. Biosynthesis of riboflavin - incorporation of D-[1-^{13}C] ribose. J. Am. Chem. Soc. 104, 3754-3755.

Bacher, A., Levan, Q., Keller, P. J., Floss, H. G., 1983. Biosynthesis of riboflavin - incorporation of ^{13}C-labeled precursors into the xylene ring. J. Biol. Chem. 258, 3431-3437.

Bacher, A., Levan, Q., Keller, P. J., Floss, H. G., 1985. Biosynthesis of riboflavin - incorporation of multiply ^{13}C-labeled precursors into the xylene ring. J. Am. Chem. Soc. 107, 6380-6385.

Bacher, A., Rieder, C., Eichinger, D., Arigoni, D., Fuchs, G., Eisenreich, W., 1998. Elucidation of novel biosynthetic pathways and metabolite flux patterns by retrobiosynthetic NMR analysis. FEMS Microbiol. Rev. 22, 567-598.

Banerjee, R., Batschauer, A., 2005. Plant blue-light receptors. Planta. 220, 498-502.

Barbau-Piednoir, E., De Keersmaecker, S. C. J., Wuyts, V., Gau, C., Pirovano, W., Costessi, A., Philipp, P., Roosens, N. H., 2015. Genome sequence of EU-unauthorized genetically modified Bacillus subtilis strain 2014-3557 overproducing riboflavin, isolated from a vitamin B2 80% feed additive. Genome Announc. 3.

Barding, G. A., Jr., Beni, S., Fukao, T., Bailey-Serres, J., Larive, C. K., 2013. Comparison of GC-MS and NMR for metabolite profiling of rice subjected to submergence stress. J. Proteome Res. 12, 898-909.

Becker, J., Klopprogge, C., Zelder, O., Heinzle, E., Wittmann, C., 2005. Amplified expression of fructose 1,6-bisphosphatase in Corynebacterium glutamicum increases in vivo flux through the pentose phosphate pathway and lysine production on different carbon sources. Appl. Environ. Microbiol. 71, 8587-8596.

Becker, J., Zelder, O., Häfner, S., Schröder, H., Wittmann, C., 2011. From zero to hero - Design-based systems metabolic engineering of *Corynebacterium glutamicum* for L-lysine production. Metab. Eng. 13, 159-168.

Blomberg, A., Adler, L., 1989. Roles of glycerol and glycerol-3-phosphate dehydrogenase (NAD+) in acquired osmotolerance of *Saccharomyces cerevisiae*. J. Bacteriol. 171, 1087-1092.

Boles, E., de Jong-Gubbels, P., Pronk, J. T., 1998. Identification and characterization of *MAE1*, the *Saccharomyces cerevisiae* structural gene encoding mitochondrial malic enzyme. J. Bacteriol. 180, 2875-82.

Bolten, C. J., Heinzle, E., Müller, R., Wittmann, C., 2009. Investigation of the central carbon metabolism of *Sorangium cellulosum*: metabolic network reconstruction and quantification of pathway fluxes. J. Microbiol. Biotechnol. 19, 23-36.

Bolten, C. J., Wittmann, C., 2008. Appropriate sampling for intracellular amino acid analysis in five phylogenetically different yeasts. Biotechnol. Lett. 30, 1993-2000.

Braus, G. H., 1991. Aromatic amino acid biosynthesis in the yeast *Saccharomyces cerevisiae*: a model system for the regulation of a eukaryotic biosynthetic pathway. Microbiol. Rev. 55, 349-370.

Bretzel, W., Schurter, W., Ludwig, B., Kupfer, E., Doswald, S., Pfister, M., van Loon, A., 1999. Commercial riboflavin production by recombinant *Bacillus subtilis*: down-stream processing and comparison of the composition of riboflavin produced by fermentation or chemical synthesis. J. Ind. Microbiol. Biotechnol. 22, 19-26.

Bücker, R., Heroven, A. K., Becker, J., Dersch, P., Wittmann, C., 2014. The pyruvate-tricarboxylic acid cycle node - a focal point of virulence control in the enteric pathogen *Yersinia pseudotuberculosis*. J. Biol. Chem. 289, 30114-30132.

Buey, R. M., Ledesma-Amaro, R., Balsera, M., de Pereda, J. M., Revuelta, J. L., 2015. Increased riboflavin production by manipulation of inosine 5'-monophosphate dehydrogenase in *Ashbya gossypii*. Appl. Microbiol. Biotechnol. 99, 9577-9589.

Burns, V. W., 1964. Regulation and coordination of purine and pyrimidine biosyntheses in yeast: I. Regulation of purine biosynthesis and its relation to transient changes in intracellular nucleotide levels. Biophys. J. 4, 151-166.

Buschke, N., Schröder, H., Wittmann, C., 2011. Metabolic engineering of *Corynebacterium glutamicum* for production of 1,5-diaminopentane from hemicellulose. Biotechnol. J. 6, 306-317.

Bykova, N. V., Stensballe, A., Egsgaard, H., Jensen, O. N., Møller, I. M., 2003. Phosphorylation of formate dehydrogenase in potato tuber mitochondria. J. Biol. Chem. 278, 26021-30.

Cain, N. E., Kaiser, C. A., 2011. Transport activity-dependent intracellular sorting of the yeast general amino acid permease. Mol. Biol. Cell. 22, 1919-1929.

Casal, M., Cardoso, H., Leão, C., 1996. Mechanisms regulating the transport of acetic acid in *Saccharomyces cerevisiae*. Microbiol. 142, 1385-1390.

Casal, M., Paiva, S., Andrade, R. P., Gancedo, C., Leão, C., 1999. The lactate-proton symport of *Saccharomyces cerevisiae* is encoded by *JEN1*. J. Bacteriol. 181, 2620-3.

Chistoserdova, L. V., Lidstrom, M. E., 1994. Genetics of the serine cycle in *Methylobacterium extorquens* AM1: Identification of *sgaA* and *mtdA* and sequences of *sgaA*, *hprA*, and *mtdA*. J. Bacteriol. 176, 1957-1968.

Christensen, B., Thykaer, J., Nielsen, J., 2000. Metabolic characterization of high- and low-yielding strains of *Penicillium chrysogenum*. Appl. Microbiol. Biotechnol. 54, 212-217.

Coquard, D., Huecas, M., Ott, M., van Dijl, J. M., van Loon, A. P. G. M., Hohmann, H. P., 1997. Molecular cloning and characterisation of the *ribC* gene from *Bacillus subtilis*: a point mutation in *ribC* results in riboflavin overproduction. Mol. Gen. Genet. 254, 81-84.

De Deken, R. H., 1966. The Crabtree effect: a regulatory system in yeast. J. Gen. Microbiol. 44, 149-56.

de Graaf, A. A., Mahle, M., Möllney, M., Wiechert, W., Stahmann, P., Sahm, H., 2000. Determination of full ^{13}C isotopomer distributions for metabolic flux analysis using heteronuclear spin echo difference NMR spectroscopy. J. Biotechnol. 77, 25-35.

Demain, A. L., 1972. Riboflavin oversynthesis. Annu. Rev. Microbiol. 26, 369-388.

Dersch, L. M., Beckers, V., Wittmann, C., 2016. Green pathways: Metabolic network analysis of plant systems. Metab. Eng. 34, 1-24.

Deutscher, J., Francke, C., Postma, P. W., 2006. How phosphotransferase system-related protein phosphorylation regulates carbohydrate metabolism in bacteria. Microbiol. Mol. Biol. Rev. 70, 939-1031.

Díaz-Fernández, D., Lozano-Martínez, P., Buey, R. M., Revuelta, J. L., Jiménez, A., 2017. Utilization of xylose by engineered strains of *Ashbya gossypii* for the production of microbial oils. Biotechnol. Biofuels. 10.

Diederichs, S., Korona, A., Staaden, A., Kroutil, W., Honda, K., Ohtake, H., Büchs, J., 2014. Phenotyping the quality of complex medium components by simple online-monitored shake flask experiments. Microb. Cell. Fact. 13.

Dietrich, F. S., Voegeli, S., Kuo, S., Philippsen, P., 2013. Genomes of *Ashbya* fungi isolated from insects reveal four mating-type loci, numerous translocations, lack of transposons, and Distinct gene duplications. G3. 3, 1225-1239.

Dmytruk, K., Lyzak, O., Yatsyshyn, V., Kluz, M., Sibirny, V., Puchalski, C., Sibirny, A., 2014. Construction and fed-batch cultivation of *Candida famata* with enhanced riboflavin production. J. Biotechnol. 172, 11-17.

DSM, 2006. Fermentative preparation of riboflavin involving recycling of lysed biomass as nutrient medium. Patent EP 1 731 617 A1, 13 Dec 2006

DSM, DSM in Grenzach - Historie. Vol. 2015, 2015.

Du, L., Su, Y., Sun, D., Zhu, W., Wang, J., Zhuang, X., Zhou, S., Lu, Y., 2008. Formic acid induces Yca1p-independent apoptosis-like cell death in the yeast *Saccharomyces cerevisiae*. FEMS Yeast Res. 8, 531-9.

Duan, Y. X., Chen, T., Chen, X., Zhao, X. M., 2010. Overexpression of glucose-6-phosphate dehydrogenase enhances riboflavin production in *Bacillus subtilis*. Appl. Microbiol. Biotechnol. 85, 1907-1914.

Eggersdorfer, M., Laudert, D., Létinois, U., McClymont, T., Medlock, J., Netscher, T., Bonrath, W., 2012. One hundred years of vitamins - a success story of the natural sciences. Angew. Chem. Int. Ed. 51, 12960-12990.

Epstein, A., Graham, G., Sklarz, W. A., 1979. Riboflavin purification. Patent US4165250, 21.08.1979

Escalera-Fanjul, X., Campero-Basaldua, C., Colón, M., González, J., Márquez, D., González, A., 2017. Evolutionary diversification of alanine transaminases in yeast: catabolic specialization and biosynthetic redundancy. Front. Microbiol. 8.

Ferreira, C., van Voorst, F., Martins, A., Neves, L., Oliveira, R., Kielland-Brandt, M. C., Lucas, C., Brandt, A., 2005. A member of the sugar transporter family, Stl1p is the glycerol/H$^+$ symporter in *Saccharomyces cerevisiae*. Mol. Biol. Cell. 16, 2068-76.

Fischer, M., Bacher, A., 2005. Biosynthesis of flavocoenzymes. Nat. Prod. Rep. 22, 324-350.

Food, Riboflavin. Dietary reference intakes: thiamin, riboflavin, niacin, vitamin B6, vitamin B12, pantothenic acid, biotin, folate and choline. The National Academies Press, Washington, DC, 1998, pp. 87-122.

Förster, C., Marienfeld, S., Wendisch, F., Krämer, R., 1998. Adaptation of the filamentous fungus *Ashbya gossypii* to hyperosmotic stress: different osmoresponse to NaCl and mannitol stress. Appl. Microbiol. Biotechnol. 50, 219-226.

Förster, C., Santos, M. A., Ruffert, S., Krämer, R., Revuelta, J. L., 1999. Physiological consequence of disruption of the *VMA1* gene in the riboflavin overproducer *Ashbya gossypii*. J. Biol. Chem. 274, 9442-9448.

Förster, J., Halbfeld, C., Zimmermann, M., Blank, L. M., 2014. A blueprint of the amino acid biosynthesis network of hemiascomycetes. FEMS Yeast Res. 14, 1090-1100.

García-Campusano, F., Anaya, V. H., Robledo-Arratia, L., Quezada, H., Hernández, H., Riego, L., González, A., 2009. *ALT1*-encoded alanine aminotransferase plays a central role in the metabolism of alanine in *Saccharomyces cerevisiae*. Can. J. Microbiol. 55, 368-74.

Geertman, J. M. A., van Dijken, J. P., Pronk, J. T., 2006. Engineering NADH metabolism in *Saccharomyces cerevisiae*: formate as an electron donor for glycerol production by anaerobic, glucose-limited chemostat cultures. FEMS Yeast Res. 6, 1193-1203.

Gey, U., Czupalla, C., Hoflack, B., Rödel, G., Krause-Buchholz, U., 2008. Yeast pyruvate dehydrogenase complex is regulated by a concerted activity of two kinases and two phosphatases. J. Biol. Chem. 283, 9759-67.

Gomes, D., Aguiar, T. Q., Dias, O., Ferreira, E. C., Domingues, L., Rocha, I., 2014. Genome-wide metabolic re-annotation of *Ashbya gossypii*: new insights into its metabolism through a comparative analysis with *Saccharomyces cerevisiae* and *Kluyveromyces lactis*. BMC Genomics. 15.

Good, N. E., Winget, G. D., Winter, W., Connolly, T. N., Izawa, S., Singh, R. M. M., 1966. Hydrogen ion buffers for biological research. Biochemistry. 5, 467-477.

Goodwin, T. W., Jones, O. T. G., 1956. Studies on the biosynthesis of riboflavin. 3. Utilization of ^{14}C-labelled serine for riboflavin biosynthesis by *Eremothecium ashbyii*. Biochem. J. 64, 9-13.

Grimmer, J., Kiefer, H., Martin, C., 1993. Method of purifying ferment-produced riboflavin. Patent US 5,210,023 A, 01 July 1991

Guillamón, J. M., van Riel, N. A., Giuseppin, M. L., Verrips, C. T., 2001. The glutamate synthase (GOGAT) of *Saccharomyces cerevisiae* plays an important role in central nitrogen metabolism. FEMS Yeast Res. 1, 169-75.

Gutiérrez-Preciado, A., Torres, A. G., Merino, E., Bonomi, H. R., Goldbaum, F. A., García-Angulo, V. A., 2015. Extensive identification of bacterial riboflavin transporters and their distribution across bacterial species. PLoS ONE. 10.

Haselbeck, R. J., McAlister-Henn, L., 1993. Function and expression of yeast mitochondiral NAD-specific and NADP-specific isocitrate dehydrogenase. J. Biol. Chem. 268, 12116-12122.

Hemberger, S., Pedrolli, D. B., Stolz, J., Vogl, C., Lehmann, M., Mack, M., 2011. RibM from *Streptomyces davawensis* is a riboflavin/roseoflavin transporter and may be useful for the optimization of riboflavin production strains. BMC Biotechnol. 11.

Higashitsuji, Y., Angerer, A., Berghaus, S., Hobl, B., Mack, M., 2007. RibR, a possible regulator of the *Bacillus subtilis* riboflavin biosynthetic operon, *in vivo* interacts with the 5'-untranslated leader of *rib* mRNA. FEMS Microbiol. Lett. 274, 48-54.

Hoffmann, T., Krug, D., Huttel, S., Müller, R., 2014. Improving natural products identification through targeted LC-MS/MS in an untargeted secondary metabolomics workflow. Anal. Chem. 86, 10780-10788.

Hohmann, H. P., Mouncey, N. J., Sauer, U., Zamboni, N., 2011. Fermentation process. Patent EP 1 481 064 B1, 20 Feb 2003

Hümbelin, M., Griesser, V., Keller, T., Schurter, W., Haiker, M., Hohmann, H. P., Ritz, H., Richter, G., Bacher, A., van Loon, A., 1999. GTP cyclohydrolase II and 3,4-dihydroxy-2-butanone 4-phosphate synthase are rate-limiting enzymes in riboflavin synthesis of an industrial *Bacillus subtilis* strain used for riboflavin production. J. Ind. Microbiol. Biotechnol. 22, 1-7.

Jahreis, K., Pimentel-Schmitt, E. F., Bruckner, R., Titgemeyer, F., 2008. Ins and outs of glucose transport systems in eubacteria. FEMS Microbiol. Rev. 32, 891-907.

Jeong, B.-Y., Wittmann, C., Kato, T., Park, E. Y., 2015. Comparative metabolic flux analysis of an *Ashbya gossypii* wild type strain and a high riboflavin-producing mutant strain. J. Biosci. Bioeng. 119, 101-106.

Jernejc, K., Legisa, M., 2002. The influence of metal ions on malic enzyme activity and lipid synthesis in *Aspergillus niger*. FEMS Microbiol. Lett. 217, 185-90.

Jiménez, A., Santos, M. A., Pompejus, M., Revuelta, J. L., 2005. Metabolic engineering of the purine pathway for riboflavin production in *Ashbya gossypii*. Appl. Microbiol. Biotechnol. 71, 5743-5751.

Jiménez, A., Santos, M. A., Revuelta, J. L., 2008. Phosphoribosyl pyrophosphate synthetase activity affects growth and riboflavin production in *Ashbya gossypii*. BMC Biotechnol. 8.

Jordà, J., Suarez, C., Carnicer, M., ten Pierick, A., Heijnen, J. J., van Gulik, W., Ferrer, P., Albiol, J., Wahl, A., 2013. Glucose-methanol co-utilization in *Pichia pastoris* studied by metabolomics and instationary ^{13}C flux analysis. BMC Syst. Biol. 7.

Kanehisa, M., Furumichi, M., Tanabe, M., Sato, Y., Morishima, K., 2017. KEGG: new perspectives on genomes, pathways, diseases and drugs. Nucleic Acids Res. 45.

Kanehisa, M., Goto, S., 2000. KEGG: kyoto encyclopedia of genes and genomes. Nucleic Acids Res. 28, 27-30.

Kanehisa, M., Sato, Y., Kawashima, M., Furumichi, M., Tanabe, M., 2016. KEGG as a reference resource for gene and protein annotation. Nucleic Acids Res. 44, 17.

Karrer, P., Schöpp, K., Benz, F., Pfaehler, K., 1935. Synthesen von Flavinen III. Helv. Chim. Acta. 18, 69-79.

Kato, T., Park, E. Y., 2006. Expression of alanine : glyoxylate aminotransferase gene from *Saccharomyces cerevisiae* in *Ashbya gossypii*. Appl. Microbiol. Biotechnol. 71, 46-52.

Kato, T., Park, E. Y., 2012. Riboflavin production by *Ashbya gossypii*. Biotechnol. Lett. 34, 611-618.

Kavitha, S., Chandra, T. S., 2009. Effect of vitamin E and menadione supplementation on riboflavin production and stress parameters in *Ashbya gossypii*. Process Biochem. 44, 934-938.

Kavitha, S., Chandra, T. S., 2014. Oxidative stress protection and glutathione metabolism in response to hydrogen peroxide and menadione in riboflavinogenic fungus *Ashbya gossypii*. Appl. Biochem. Biotechnol. 174, 2307-2325.

Kind, S., Kreye, S., Wittmann, C., 2011. Metabolic engineering of cellular transport for overproduction of the platform chemical 1,5-diaminopentane in *Corynebacterium glutamicum*. Metab. Eng. 13, 617-627.

Kirchner, F., Mauch, K., Schmid, J., 2014. Process for the production of riboflavin. Patent US 8,759,024 B2, 24 June 2014

Kleijn, R. J., Geertman, J.-M. A., Nfor, B. K., Ras, C., Schipper, D., Pronk, J. T., Heijnen, J. J., van Maris, A. J. A., van Winden, W. A., 2007. Metabolic flux analysis of a glycerol-overproducing *Saccharomyces cerevisiae* strain based on GC-MS, LC-MS and NMR-derived ^{13}C-labelling data. FEMS Yeast Res. 7, 216-231.

Klein, M., Swinnen, S., Thevelein, J. M., Nevoigt, E., 2017. Glycerol metabolism and transport in yeast and fungi: established knowledge and ambiguities. Environ. Microbiol. 19, 878-893.

Knorr, B., Schlieker, H., Hohmann, H.-P., Weuster-Botz, D., 2007. Scale-down and parallel operation of the riboflavin production process with *Bacillus subtilis*. Chem. Eng. J. Bioch. Eng. 33, 263-274.

Kohlstedt, M., Becker, J., Wittmann, C., 2010. Metabolic fluxes and beyond-systems biology understanding and engineering of microbial metabolism. Appl. Microbiol. Biotechnol. 88, 1065-1075.

Kohlstedt, M., Sappa, P. K., Meyer, H., Maaß, S., Zaprasis, A., Hoffmann, T., Becker, J., Steil, L., Hecker, M., van Dijl, J. M., Lalk, M., Mäder, U., Stülke, J., Bremer, E., Völker, U., Wittmann, C., 2014. Adaptation of *Bacillus subtilis* carbon core metabolism to simultaneous nutrient limitation and osmotic challenge: a multi-omics perspective. Environ. Microbiol. 16, 1898-1917.

Kojima, H., Ogawa, Y., Kawamura, K. S., 2000. Method of producing L-lysine by fermentation. Patent US 6,040,160, 21 Mar 2000

Krömer, J. O., Fritz, M., Heinzle, E., Wittmann, C., 2005. In vivo quantification of intracellular amino acids and intermediates of the methionine pathway in *Corynebacterium glutamicum*. Anal. Biochem. 340, 171-173.

Krömer, J. O., Sorgenfrei, O., Klopprogge, K., Heinzle, E., Wittmann, C., 2004. In-depth profiling of lysine-producing *Corynebacterium glutamicum* by combined analysis of the transcriptome, metabolome, and fluxome. J. Bacteriol. 186, 1769-1784.

Kuhn, R., 1936. Lactoflavin (Vitamin B_2). Angew. Chem. 49, 6-10.

Kuhn, R., György, P., Wagner-Jauregg, T., 1933a. Über Lactoflavin, den Farbstoff der Molke. Ber. Dtsch. Chem. Ges. 66, 1034-1038.

Kuhn, R., György, P., Wagner-Jauregg, T., 1933b. Über Ovoflavin, den Farbstoff des Eiklars. Ber. Dtsch. Chem. Ges. 66, 576-580.

Lacroute, F., 1968. Regulation of pyrimidine biosynthesis in *Saccharomyces cerevisiae*. J. Bacteriol. 95, 824-32.

Laine, R. A., Sweeley, C. C., 1971. Analysis of trimethylsilyl O-methyloximes of carbohydrates by combined gas-liquid chromatography-mass spectrometry. Anal. Biochem. 43, 533-8.

Lange, A., Becker, J., Schulze, D., Cahoreau, E., Portais, J.-C., Haefner, S., Schröder, H., Krawczyk, J., Zelder, O., Wittmann, C., 2017. Bio-based succinate from sucrose: High-resolution ^{13}C metabolic flux analysis and metabolic engineering of the rumen bacterium *Basfia succiniciproducens*. Metab. Eng. 44, 198-212.

Large, V., Brunengraber, H., Odeon, M., Beylot, M., 1997. Use of labeling pattern of liver glutamate to calculate rates of citric acid cycle and gluconeogenesis. Am. J. Physiol. Endocrinol. Metab. 272, E51-E58.

Lastauskienė, E., Zinkevičienė, A., Girkontaitė, I., Kaunietis, A., Kvedarienė, V., 2014. Formic acid and acetic acid induce a programmed cell death in pathogenic Candida species. Curr. Microbiol. 69, 303-310.

Ledesma-Amaro, R., Buey, R. M., Revuelta, J. L., 2015a. Increased production of inosine and guanosine by means of metabolic engineering of the purine pathway in *Ashbya gossypii*. Microb. Cell. Fact. 14.

Ledesma-Amaro, R., Buey, R. M., Revuelta, J. L., 2016. The filamentous fungus *Ashbya gossypii* as a competitive industrial inosine producer. Biotechnol. Bioeng. 113, 2060-2063.

Ledesma-Amaro, R., Kerkhoven, E. J., Luis Revuelta, J., Nielsen, J., 2014a. Genome scale metabolic modeling of the riboflavin overproducer *Ashbya gossypii*. Biotechnol. Bioeng. 111, 1191-1199.

Ledesma-Amaro, R., Lozano-Martínez, P., Jiménez, A., Luis Revuelta, J., 2015b. Engineering *Ashbya gossypii* for efficient biolipid production. Bioengineered. 6, 119-123.

Ledesma-Amaro, R., Santos, M. A., Jiménez, A., Luis Revuelta, J., 2014b. Strain design of *Ashbya gossypii* for single-cell oil production. Appl. Environ. Microbiol. 80, 1237-1244.

Ledesma-Amaro, R., Serrano-Amatriain, C., Jiménez, A., Revuelta, J. L., 2015c. Metabolic engineering of riboflavin production in *Ashbya gossypii* through pathway optimization. Microb. Cell Fact. 14, 163-163.

Lee, K., Park, Y., Han, J., Park, J., Choi, H., 2004. Microorganism for producing riboflavin and method for producing riboflavin using the same. Patent US2004/0110249A1, 10.06.2004

Lee, Y. J., Jang, J. W., Kim, K. J., Maeng, P. J., 2011. TCA cycle-independent acetate metabolism via the glyoxylate cycle in *Saccharomyces cerevisiae*. Yeast. 28, 153-166.

Liepman, A. H., Olsen, L. J., 2001. Peroxisomal alanine : glyoxylate aminotransferase (AGT1) is a photorespiratory enzyme with multiple substrates in *Arabidopsis thaliana*. Plant J. 25, 487-498.

Liepman, A. H., Olsen, L. J., 2003. Alanine aminotransferase homologs catalyze the glutamate:glyoxylate aminotransferase reaction in peroxisomes of Arabidopsis. Plant Physiol. 131, 215-227.

Lim, S. H., Choi, J. S., Park, E. Y., 2001. Microbial production of riboflavin using riboflavin overproducers, *Ashbya gossypii*, *Bacillus subtilis*, and *Candida famate*: An overview. Biotechnol. Bioprocess Eng. 6, 75-88.

Lin, Z., Li, W.-H., 2011. Expansion of hexose transporter genes was associated with the evolution of aerobic fermentation in yeasts. Mol. Biol. Evol. 28, 131-142.

Lin, Z. Q., Xu, Z. B., Li, Y. F., Wang, Z. W., Chen, T., Zhao, X. M., 2014. Metabolic engineering of *Escherichia coli* for the production of riboflavin. Microb. Cell Fact. 13.

Liu, Y., Ding, M. Z., Ling, W., Yang, Y., Zhou, X., Li, B. Z., Chen, T., Nie, Y., Wang, M. X., Zeng, B. X., Li, X., Liu, H., Sun, B. D., Xu, H. M., Zhang, J. M., Jiao, Y., Hou, Y. A., Yang, H., Xiao, S. J., Lin, Q. C., He, X. Z., Liao, W. J., Jin, Z. Q., Xie, Y. F., Zhang, B. F., Li, T. Y., Lu, X., Li, J. B., Zhang, F., Wu, X. L., Song, H., Yuan, Y. J., 2017. A three-species microbial consortium for power generation. Energy Environ. Sci. 10, 1600-1609.

Ljungdahl, P. O., Daignan-Fornier, B., 2012. Regulation of amino acid, nucleotide, and phosphate metabolism in *Saccharomyces cerevisiae*. Genetics. 190, 885-929.

Lozano-Martínez, P., Buey, R. M., Ledesma-Amaro, R., Jiménez, A., Revuelta, J. L., 2016. Engineering *Ashbya gossypii* strains for *de novo* lipid production using industrial by-products. Microb. Biotechnol. 10, 425-433.

Mack, M., van Loon, A., Hohmann, H. P., 1998. Regulation of riboflavin biosynthesis in *Bacillus subtilis* is affected by the activity of the flavokinase/flavin adenine dinucleotide synthetase encoded by *ribC*. J. Bacteriol. 180, 950-955.

Maeting, I., Schmidt, G., Sahm, H., Stahmann, K. P., 2000. Role of a peroxisomal NADP-specific isocitrate dehydrogenase in the metabolism of the riboflavin overproducer *Ashbya gossypii*. J. Mol. Catal. B Enzym. 10, 335-343.

Magasanik, B., 2003. Ammonia assimilation by *Saccharomyces cerevisiae*. Eukaryot. Cell. 2, 827-829.

Magnusson, I., Schumann, W. C., Bartsch, G. E., Chandramouli, V., Kumaran, K., Wahren, J., Landau, B. R., 1991. Noninvasive tracing of Krebs cycle metabolism in liver. J. Biol. Chem. 266, 6975-84.

Malzahn, R. C., Phillips, R. F., Hanson, A. M., 1959. Riboflavin process. Patent US 2,876,169, 03 Mar 1959

Man, Z.-w., Rao, Z.-m., Cheng, Y.-p., Yang, T.-w., Zhang, X., Xu, M.-j., Xu, Z.-h., 2014. Enhanced riboflavin production by recombinant *Bacillus subtilis* RF1 through the optimization of agitation speed. World J. Microb. Biot. 30, 661-667.

Markley, J. L., Brüschweiler, R., Edison, A. S., Eghbalnia, H. R., Powers, R., Raftery, D., Wishart, D. S., 2017. The future of NMR-based metabolomics. Curr. Opin. Biotechnol. 43, 34-40.

Martínez-Blanco, H., Reglero, A., Fernández-Valverde, M., Ferrero, M. A., Moreño, M. A., Penalva, M. A., Luengo, J. M., 1992. Isolation and characterization of the acetyl-CoA synthetase from Penicillium chrysogenum - involvement of this enzyme in the biosynthesis of penicillins. J. Biol. Chem. 267, 5474-5481.

Mashego, M. R., Rumbold, K., De Mey, M., Vandamme, E., Soetaert, W., Heijnen, J. J., 2007. Microbial metabolomics: past, present and future methodologies. Biotechnol. Lett. 29, 1-16.

Massey, V., 2000. The chemical and biological versatility of riboflavin. Biochem. Soc. Trans. 28, 283-296.

Mateos, L., Jiménez, A., Revuelta, J. L., Santos, M. A., 2006. Purine biosynthesis, riboflavin production, and trophic-phase span are controlled by a myb-related transcription factor in the fungus Ashbya gossypii. Appl. Environ. Microbiol. 72, 5052-5060.

Matsui, H., Sato, K., Enei, H., Hirose, Y., 1977. Mutation of an inosine-producing strain of Bacillus subtilis to DL-methionine sulfoxide resistance for guanosine production. Appl. Environ. Microbiol. 34, 337-341.

Matsui, H., Sato, K., Enei, H., Hirose, Y., 1979. Production of guanosine by psicofuranine and decoyinine resistant mutants of Bacillus subtilis. Agric. Biol. Chem. 43, 1739-1744.

Matsuzawa, T., Ohashi, T., Hosomi, A., Tanaka, N., Tohda, H., Takegawa, K., 2010. The $gld1^+$ gene encoding glycerol dehydrogenase is required for glycerol metabolism in Schizosaccharomyces pombe. Appl. Microbiol. Biotechnol. 87, 715-27.

Merico, A., Sulo, P., Piskur, J., Compagno, C., 2007. Fermentative lifestyle in yeasts belonging to the Saccharomyces complex. FEBS J. 274, 976-89.

Messenguy, F., Colin, D., Tenhave, J. P., 1980. Regulation of compartmentation of amino acid pools in Saccharomyces cerevisiae and its effects on metabolic control. Eur. J. Biochem. 108, 439-447.

Mickelson, M. N., 1950. The metabolism of glucose by Ashbya gossypii. J. Bacteriol. 59, 659-666.

Mickelson, M. N., Schuler, M. N., 1953. Oxidation of acetate by Ashbya gossypii. J. Bacteriol. 65, 297-304.

Ming, H., Pizarro, A. V. L., Park, E. Y., 2003. Application of waste activated bleaching earth containing rapeseed oil on riboflavin production in the culture of Ashbya gossypii. Biotechnol. Progr. 19, 410-417.

Mironov, A. S., Gusarov, I., Rafikov, R., Lopez, L. E., Shatalin, K., Kreneva, R. A., Perumov, D. A., Nudler, E., 2002. Sensing small molecules by nascent RNA: A mechanism to control transcription in bacteria. Cell. 111, 747-756.

Mitra, S., Thawrani, D., Banerjee, P., Gachhui, R., Mukherjee, J., 2012. Induced biofilm cultivation enhances riboflavin production by an intertidally derived Candida famata. Appl. Biochem. Biotechnol. 166, 1991-2006.

Miyamoto, Y., Sancar, A., 1998. Vitamin B_2-based blue-light photoreceptors in the retinohypothalamic tract as the photoactive pigments for setting the circadian clock in mammals. PNAS. 95, 6097-6102.

Monschau, N., Sahm, H., Stahmann, K. P., 1998. Threonine aldolase overexpression plus threonine supplementation enhanced riboflavin production in Ashbya gossypii. Appl. Environ. Microbiol. 64, 4283-4290.

Morin, P. J., Subramanian, G. S., Gilmore, T. D., 1992. AAT1, a gene encoding a mitochondrial aspartate aminotransferase in Saccharomyces cerevisiae. BBA - Gene Structure and Expression. 1171, 211-214.

Murakami, K., Yoshino, M., 1997. Inactivation of aconitase in yeast exposed to oxidative stress. IUBMB Life. 41, 481-486.

Murphy, Michael P., 2009. How mitochondria produce reactive oxygen species. Biochem. J. 417, 1-13.

Nevoigt, E., Stahl, U., 1997. Osmoregulation and glycerol metabolism in the yeast *Saccharomyces cerevisiae*. FEMS Microbiol. Rev. 21, 231-241.

Nieland, S., Stahmann, K. P., 2013. A developmental stage of hyphal cells shows riboflavin overproduction instead of sporulation in *Ashbya gossypii*. Appl. Microbiol. Biotechnol. 97, 10143-10153.

Nisbet, B. A., Slaughter, J. C., 1980. Glutamate dehydrogenase and glutamate synthase from the yeast *Kluyveromyces fragilis*: variability in occurrence and properties. FEMS Microbiol. Lett. 7, 319-321.

Northrop-Clewes, C. A., Thurnham, D. I., 2012. The discovery and characterization of riboflavin. Ann. Nutr. Metab. 61, 224-230.

O'Neil, M. J., 2006. Riboflavin. In: O'Neil, M. J., (Ed.), The Merck Index - Encyclopedia of chemicals, drugs & biologicals, vol. 14. Whitehouse Station, New Jersey, USA, pp. 1413.

Outten, C. E., Culotta, V. C., 2003. A novel NADH kinase is the mitochondrial source of NADPH in *Saccharomyces cerevisiae*. EMBO J. 22, 2015-2024.

Özcan, S., Johnston, M., 1999. Function and regulation of yeast hexose transporters. Microbiol. Mol. Biol. Rev. 63, 554-569.

Paiva, S., Devaux, F., Barbosa, S., Jacq, C., Casal, M., 2004. Ady2p is essential for the acetate permease activity in the yeast *Saccharomyces cerevisiae*. Yeast. 21, 201-210.

Pan, Z., Raftery, D., 2007. Comparing and combining NMR spectroscopy and mass spectrometry in metabolomics. Anal. Bioanal. Chem. 387, 525-527.

Panagiotou, G., Grotkjær, T., Hofmann, G., Bapat, P. M., Olsson, L., 2009. Overexpression of a novel endogenous NADH kinase in *Aspergillus nidulans* enhances growth. Metab. Eng. 11, 31-39.

Papagianni, M., 2004. Fungal morphology and metabolite production in submerged mycelial processes. Biotechnol. Adv. 22, 189-259.

Paracchini, V., Petrillo, M., Reiting, R., Angers-Loustau, A., Wahler, D., Stolz, A., Schönig, B., Matthies, A., Bendiek, J., Meinel, D. M., Pecoraro, S., Busch, U., Patak, A., Kreysa, J., Grohmann, L., 2017. Molecular characterization of an unauthorized genetically modified *Bacillus subtilis* production strain identified in a vitamin B_2 feed additive. Food Chem. 230, 681-689.

Park, E. Y., Ito, Y., Nariyama, M., Sugimoto, T., Lies, D., Kato, T., 2011. The improvement of riboflavin production in *Ashbya gossypii* via disparity mutagenesis and DNA microarray analysis. Appl. Microbiol. Biotechnol. 91, 1315-1326.

Park, E. Y., Kato, A., Ming, H., 2004. Utilization of waste activated bleaching earth containing palm oil in riboflavin production by *Ashbya gossypii*. J. Am. Oil Chem. Soc. 81, 57-62.

Park, E. Y., Ming, H., 2004. Oxidation of rapeseed oil in waste activated bleaching earth and its effect on riboflavin production in culture of *Ashbya gossypii*. J. Biosci. Bioeng. 97, 59-64.

Park, S. M., Shaw-Reid, C., Sinskey, A. J., Stephanopoulos, G., 1997. Elucidation of anaplerotic pathways in *Corynebacterium glutamicum* via ^{13}C NMR spectroscopy and GC-MS. Appl. Microbiol. Biotechnol. 47, 430-440.

Pasternack, L. B., Littlepage, L. E., Laude, D. A., Appling, D. R., 1996. ^{13}C NMR analysis of the use of alternative donors to the tetrahydrofolate-dependent one-carbon pools in *Saccharomyces cerevisiae*. Arch. Biochem. Biophys. 326, 158-165.

Pedrolli, D. B., Kühm, C., Sévin, D. C., Vockenhuber, M. P., Sauer, U., Suess, B., Mack, M., 2015. A dual control mechanism synchronizes riboflavin and sulphur metabolism in *Bacillus subtilis*. PNAS. 112, 14054-9.

Perkins, J. B., Sloma, A., Hermann, T., Theriault, K., Zachgo, E., Erdenberger, T., Hannett, N., Chatterjee, N. P., Williams, V., Rufo, G. A., Hatch, R., Pero, J., 1999. Genetic engineering of *Bacillus subtilis* for the commercial production of riboflavin. J. Ind. Microbiol. Biotechnol. 22, 8-18.

Piper, M. D., Hong, S. P., Ball, G. E., Dawes, I. W., 2000. Regulation of the balance of one-carbon metabolism in *Saccharomyces cerevisiae*. J. Biol. Chem. 275, 30987-30995.

Piper, M. D. W., Hong, S.-P., Eißing, T., Sealey, P., Dawes, I. W., 2002. Regulation of the yeast glycine cleavage genes is responsive to the availability of multiple nutrients. FEMS Yeast Res. 2, 59-71.

Plaut, G. W. E., 1954a. Biosynthesis of riboflavin. 1. Incorporation of ^{14}C-labeled compounds into ring B and ring C. J. Biol. Chem. 208, 513-520.

Plaut, G. W. E., 1954b. Biosynthesis of riboflavin. 2. Incorporation of ^{14}C-labeled compounds into ring A. J. Biol. Chem. 211, 111-116.

Plaut, G. W. E., Broberg, P. L., 1956. Biosynthesis of riboflavin. 3. Incorporation of ^{14}C-labeled compounds into the ribityl side chain. J. Biol. Chem. 219, 131-138.

Posch, A. E., Spadiut, O., Herwig, C., 2012. Switching industrial production processes from complex to defined media: method development and case study using the example of *Penicillium chrysogenum*. Microb. Cell Fact. 11, 88-88.

Pridham, T. G., Raper, K. B., 1950. *Ashbya gossypii* - its significance in nature and in the laboratory. Mycologia. 42, 603-623.

Quek, L.-E., Wittmann, C., Nielsen, L. K., Krömer, J. O., 2009. OpenFLUX: efficient modelling software for ^{13}C-based metabolic flux analysis. Microb. Cell Fact. 8.

Ravasio, D., Wendland, J., Walther, A., 2014. Major contribution of the Ehrlich pathway for 2-phenylethanol/rose flavor production in *Ashbya gossypii*. FEMS Yeast Res. 14, 833-844.

Revuelta, J. L., Ledesma-Amaro, R., Lozano-Martinez, P., Díaz-Fernández, D., Buey, R. M., Jiménez, A., 2016. Bioproduction of riboflavin: a bright yellow history. J. Ind. Microbiol. Biotechnol. 44, 659-665.

Ribeiro, O., Domingues, L., Penttilä, M., Wiebe, M. G., 2012. Nutritional requirements and strain heterogeneity in *Ashbya gossypii*. J. Basic Microbiol. 52, 582-589.

Ribeiro, O., Wiebe, M., Ilmen, M., Domingues, L., Penttilä, M., 2010. Expression of *Trichoderma reesei* cellulases CBHI and EGI in *Ashbya gossypii*. Appl. Microbiol. Biotechnol. 87, 1437-1446.

Richter, G., Fischer, M., Krieger, C., Eberhardt, S., Luttgen, H., Gerstenschlager, I., Bacher, A., 1997. Biosynthesis of riboflavin: Characterization of the bifunctional deaminase-reductase of *Escherichia coli* and *Bacillus subtilis*. J. Bacteriol. 179, 2022-2028.

Rolfes, R. J., 2006. Regulation of purine nucleotide biosynthesis: in yeast and beyond. Biochem. Soc. Trans. 34, 786-90.

Ryffel, F., Helfrich, E. J. N., Kiefer, P., Peyriga, L., Portais, J. C., Piel, J., Vorholt, J. A., 2016. Metabolic footprint of epiphytic bacteria on *Arabidopsis thaliana* leaves. ISME J. 10, 632-643.

Sahm, H., Antranikian, G., Stahmann, K.-P., Takors, R., 2013. Riboflavin (Vitamin B2). In: Sahm, H., Antranikian, G., Stahmann, K.-P., Takors, R., Eds.), Industrielle Mikrobiologie. Springer-Verlag, Berlin-Heidelberg, pp. 132-140.

Saksinchai, S., Suphantharika, M., Verduyn, C., 2001. Application of a simple yeast extract from spent brewer's yeast for growth and sporulation of *Bacillus thuringiensis* subsp. *kurstaki*: a physiological study. World J. Microbiol. Biotechnol. 17, 307-316.

Satapati, S., Sunny, N. E., Kucejova, B., Fu, X., He, T. T., Méndez-Lucas, A., Shelton, J. M., Perales, J. C., Browning, J. D., Burgess, S. C., 2012. Elevated TCA cycle function in the pathology of diet-induced hepatic insulin resistance and fatty liver. J. Lipid Res. 53, 1080-1092.

Schatschneider, S., Huber, C., Neuweger, H., Watt, T. F., Puhler, A., Eisenreich, W., Wittmann, C., Niehaus, K., Vorholter, F. J., 2014. Metabolic flux pattern of glucose utilization by *Xanthomonas campestris* pv. campestris: prevalent role of the Entner-Doudoroff pathway and minor fluxes through the pentose phosphate pathway and glycolysis. Mol. Biosyst. 10, 2663-2676.

Schlösser, T., Schmidt, G., Stahmann, K. P., 2001. Transcriptional regulation of 3,4-dihydroxy-2-butanone 4-phosphate synthase. Microbiol. SGM. 147, 3377-3386.

Schlösser, T., Wiesenburg, A., Gätgens, C., Funke, A., Viets, U., Vijayalakshmi, S., Nieland, S., Stahmann, K. P., 2007. Growth stress triggers riboflavin overproduction in *Ashbya gossypii*. Appl. Microbiol. Biotechnol. 76, 569-578.

Schlüpen, C., Santos, M. A., Weber, U., de Graaf, A., Revuelta, J. L., Stahmann, K. P., 2003. Disruption of the *SHM2* gene, encoding one of two serine hydroxymethyltransferase isoenzymes, reduces the flux from glycine to serine in *Ashbya gossypii*. Biochem. J. 369, 263-273.

Schmidt, G., Stahmann, K. P., Kaesler, B., Sahm, H., 1996a. Correlation of isocitrate lyase activity and riboflavin formation in the riboflavin overproducer *Ashbya gossypii*. Microbiol. 142, 419-426.

Schmidt, G., Stahmann, K. P., Sahm, H., 1996b. Inhibition of purified isocitrate lyase identified itaconate and oxalate as potential antimetabolites for the riboflavin. Microbiol. 142, 411-417.

Schwechheimer, S. K., Park, E. Y., Revuelta, J. L., Becker, J., Wittmann, C., 2016. Biotechnology of riboflavin. Appl. Microbiol. Biotechnol. 100, 2107-2119.

Serrano-Amatriain, C., Ledesma-Amaro, R., López-Nicolás, R., Ros, G., Jiménez, A., Revuelta, J. L., 2016. Folic acid production by engineered *Ashbya gossypii*. Metab. Eng. 38, 473-482.

Shen, Y.-Q., Burger, G., 2009. Plasticity of a key metabolic pathway in fungi. Funct. Integr. Genomics. 9, 145-151.

Shi, S. B., Chen, T., Zhang, Z. G., Chen, X., Zhao, X. M., 2009a. Transcriptome analysis guided metabolic engineering of *Bacillus subtilis* for riboflavin production. Metab. Eng. 11, 243-252.

Shi, S. B., Shen, Z., Chen, X., Chen, T., Zhao, X. M., 2009b. Increased production of riboflavin by metabolic engineering of the purine pathway in *Bacillus subtilis*. Biochem. Eng. J. 46, 28-33.

Shi, T., Wang, Y. C., Wang, Z. W., Wang, G. L., Liu, D. Y., Fu, J., Chen, T., Zhao, X. M., 2014. Deregulation of purine pathway in *Bacillus subtilis* and its use in riboflavin biosynthesis. Microb. Cell. Fact. 13.

Silva, R., Aguiar, T. Q., Domingues, L., 2015. Blockage of the pyrimidine biosynthetic pathway affects riboflavin production in *Ashbya gossypii*. J. Biotechnol. 193, 37-40.

Snell, K., 1984. Enzymes of serine metabolism in normal, developing and neoplastic rat tissues. Adv. Enzyme Regul. 22, 325-400.

Sørensen, J. L., Sondergaard, T. E., 2014. The effects of different yeast extracts on secondary metabolite production in *Fusarium*. Int. J. Food Microbiol. 170, 55-60.

Stahmann, K. P., Arst, H. N., Althöfer, H., Revuelta, J. L., Monschau, N., Schlüpen, C., Gätgens, C., Wisenburg, A., Schlösser, T., 2001. Riboflavin, overproduced during sporulation of *Ashbya gossypii*, protects its hyaline spores against ultraviolet light. Environ. Microbiol. 3, 545-550.

Stahmann, K. P., Böddecker, T., Sahm, H., 1997. Regulation and properties of a fungal lipase showing interfacial inactivation by gas bubbles, or droplets of lipid or fatty acid. Eur. J. Biochem. 244, 220-225.

Stahmann, K. P., Kupp, C., Feldmann, S. D., Sahm, H., 1994. Formation and degradation of lipid bodies found in the riboflavin-producing fungus *Ashbya gossypii*. Appl. Microbiol. Biotechnol. 42, 121-127.

Stahmann, K. P., Revuelta, J. L., Seulberger, H., 2000. Three biotechnical processes using *Ashbya gossypii*, *Candida famata*, or *Bacillus subtilis* compete with chemical riboflavin production. Appl. Microbiol. Biotechnol. 53, 509-516.

Storhas, W., Metz, R., 2006. Riboflavin - Vitamin B₂. In: Heinzle, E., Biwer, A., Cooney, C., Eds.), Development of sustainable bioprocesses: modeling and assessment. John Wiley & Sons, Ltd., New Jersey, USA, pp. 167-177.

Sugimoto, T., Kanamasa, S., Kato, T., Park, E. Y., 2009. Importance of malate synthase in the glyoxylate cycle of Ashbya gossypii for the efficient production of riboflavin. Appl. Microbiol. Biotechnol. 83, 529-539.

Sugimoto, T., Kato, T., Park, E. Y., 2014. Functional analysis of cis-aconitate decarboxylase and trans-aconitate-metabolism in riboflavin-producing filamentous Ashbya gossypii. J. Biosci. Bioeng. 117, 563-568.

Sugimoto, T., Morimoto, A., Nariyama, M., Kato, T., Park, E. Y., 2010. Isolation of an oxalate-resistant Ashbya gossypii strain and its improved riboflavin production. J. Ind. Microbiol. Biotechnol. 37, 57-64.

Szyperski, T., 1995. Biosynthetically directed fractional ¹³C-labeling of proteinogenic amino acids. An efficient analytical tool to investigate intermediary metabolism. Eur. J. Biochem. 232, 433-48.

Taniguchi, H., Wendisch, V. F., 2015. Exploring the role of sigma factor gene expression on production by Corynebacterium glutamicum: sigma factor H and FMN as example. Front. Microbiol. 6.

Tanner, F. W., Vojnovich, C., Vanlanen, J. M., 1949. Factors affecting riboflavin production by Ashbya gossypii. J. Bacteriol. 58, 737-745.

Tanner, J. F. W., Wickerham, L. J., Van Lanen, J. M., 1948. Biological process for the production of riboflavin. Patent US 2,445,128, 13.07.1948

Tännler, S., Decasper, S., Sauer, U., 2008a. Maintenance metabolism and carbon fluxes in Bacillus species. Microb. Cell Fact. 7, 19-19.

Tännler, S., Fischer, E., Le Coq, D., Doan, T., Jamet, E., Sauer, U., Aymerich, S., 2008b. CcpN controls central carbon fluxes in Bacillus subtilis. 190, 6178-6187.

Tom, G. D., Viswanath-Reddy, M., Howe, H. B., Jr., 1978. Effect of carbon source on enzymes involved in glycerol metabolism in Neurospora crassa. Arch. Microbiol. 117, 259-63.

van den Berg, M. A., de Jong-Gubbels, P., Kortland, C. J., van Dijken, J. P., Pronk, J. T., Steensma, H. Y., 1996. The two acetyl-coenzyme A synthetases of Saccharomyces cerevisiae differ with respect to kinetic properties and transcriptional regulation. J. Biol. Chem. 271, 28953-28959.

van Winden, W. A., Wittmann, C., Heinzle, E., Heijnen, J. J., 2002. Correcting mass isotopomer distributions for naturally occurring isotopes. Biotechnol. Bioeng. 80, 477-479.

Vanoni, M. A., Curti, B., 2008. Structure–function studies of glutamate synthases: A class of self-regulated iron-sulfur flavoenzymes essential for nitrogen assimilation. IUBMB Life. 60, 287-300.

Vorapreeda, T., Thammarongtham, C., Cheevadhanarak, S., Laoteng, K., 2012. Alternative routes of acetyl-CoA synthesis identified by comparative genomic analysis: involvement in the lipid production of oleaginous yeast and fungi. Microbiol. 158, 217-228.

Walther, A., Wendland, J., 2012. Yap1-dependent oxidative stress response provides a link to riboflavin production in Ashbya gossypii. Fungal Genet. Biol. 49, 697-707.

Wang, G. L., Bai, L., Wang, Z. W., Shi, T., Chen, T., Zhao, X. M., 2014. Enhancement of riboflavin production by deregulating gluconeogenesis in Bacillus subtilis. World J. Microbiol. Biotechnol. 30, 1893-1900.

Wendland, J., Dünkler, A., Walther, A., 2011. Characterization of α-factor pheromone and pheromone receptor genes of Ashbya gossypii. FEMS Yeast Res. 11, 418-429.

Wendland, J., Walther, A., 2005. Ashbya gossypii: A model for fungal developmental biology. Nat. Rev. Microbiol. 3, 421-429.

Wiechert, W., 2001. ¹³C metabolic flux analysis. Metab. Eng. 3, 195-206.

Wiechert, W., Mollney, M., Petersen, S., de Graaf, A. A., 2001. A universal framework for ^{13}C metabolic flux analysis. Metab. Eng. 3, 265-83.

Wipf, D., Ludewig, U., Tegeder, M., Rentsch, D., Koch, W., Frommer, W. B., 2002. Conservation of amino acid transporters in fungi, plants and animals. Trends Biochem. Sci. 27, 139-147.

Wittmann, C., 2007. Fluxome analysis using GC-MS. Microb. Cell. Fact. 6, 17.

Wittmann, C., Hans, M., Heinzle, E., 2002. In vivo analysis of intracellular amino acid labelings by GC/MS. Anal. Biochem. 307, 379-382.

Wittmann, C., Heinzle, E., 1999. Mass spectrometry for metabolic flux analysis. Biotechnol. Bioeng. 62, 739-750.

Wittmann, C., Heinzle, E., 2001. Modeling and experimental design for metabolic flux analysis of lysine-producing Corynebacteria by mass spectrometry. Metab. Eng. 3, 173-191.

Wittmann, C., Heinzle, E., 2002. Genealogy profiling through strain improvement by using metabolic network analysis: Metabolic flux genealogy of several generations of lysine-producing Corynebacteria. Appl. Environ. Microbiol. 68, 5843-5859.

Wittmann, C., Heinzle, E., 2005. Metabolic activity profiling by ^{13}C tracer experiments and mass spectrometry in Corynebacterium glutamicum. In: Barredo, J.-L., (Ed.), Microbial processes and products. Humana Press, Totowa, NJ, pp. 191-204.

Wolf, R., Reiff, F., Wittmann, R., Butzke, J., 1983. Verfahren zur Herstellung von Riboflavin. Patent EP 0 020 959 B1, 29 June 1983

Wood, H. G., 1968. Mechanism of formation of oxaloacetate and phosphoenol pyruvate from pyruvate. J. Vitaminol. S 14, 59-67.

Yagi, T., Kagamiyama, H., Nozaki, M., 1982. Aspartate:2-oxoglutarate aminotransferase from bakers' yeast: crystallization and characterization. J. Biochem. 92, 35-43.

Yakimov, A. P., Seregina, T. A., Kholodnyak, A. A., Kreneva, R. A., Mironov, A. S., Perumov, D. A., Timkovskii, A. L., 2014. Possible function of the ribT gene of Bacillus subtilis: theoretical prediction, cloning, and expression. Acta Naturae. 6, 106-109.

Zamboni, N., Fischer, E., Sauer, U., 2005. FiatFlux - a software for metabolic flux analysis from ^{13}C-glucose experiments. BMC Bioinform. 6.

Zamboni, N., Maaheimo, H., Szyperski, T., Hohmann, H. P., Sauer, U., 2004. The phosphoenolpyruvate carboxykinase also catalyzes C_3 carboxylation at the interface of glycolysis and the TCA cycle of Bacillus subtilis. Metab. Eng. 6, 277-284.

Zamboni, N., Mouncey, N., Hohmann, H. P., Sauer, U., 2003. Reducing maintenance metabolism by metabolic engineering of respiration improves riboflavin production by Bacillus subtilis. Metab. Eng. 5, 49-55.

Zelle, R. M., Trueheart, J., Harrison, J. C., Pronk, J. T., van Maris, A. J. A., 2010. Phosphoenolpyruvate carboxykinase as the sole anaplerotic enzyme in Saccharomyces cerevisiae. Appl. Environ. Microbiol. 76, 5383-5389.

Zhang, J. Y., Reddy, J., Buckland, B., Greasham, R., 2003. Toward consistent and productive complex media for industrial fermentations: Studies on yeast extract for a recombinant yeast fermentation process. Biotechnol. Bioeng. 82, 640-652.

Zhu, Y., Chen, X., Chen, T., Shi, S., Zhao, X., 2006. Over-expression of glucose dehydrogenase improves cell growth and riboflavin production in Bacillus subtilis. Biotechnol. Lett. 28, 1667-1672.

* 9 7 8 3 7 3 6 9 9 8 8 1 0 *